To Barbara

3.25

processes of organic evolution

PRENTICE-HALL BIOLOGICAL SCIENCE SERIES

William D. McElroy and Carl P. Swanson, *Editors*

*Biochemical Systematics,** RALPH E. ALSTON AND B. L. TURNER
The Cellular Functions of Membrane Transport, JOSEPH F. HOFFMAN, ED.
Classic Papers in Genetics, JAMES A. PETERS
Experimental Biology, RICHARD W. VAN NORMAN
Foundations of Experimental Embryology, BENJAMIN H. WILLIER AND JANE M. OPPENHEIMER
General and Comparative Physiology, WILLIAM S. HOAR
Mechanisms of Body Functions, DEXTER M. EASTON
Milestones in Microbiology, THOMAS D. BROCK
Papers on Human Genetics, SAMUEL H. BOYER, IV
Poisonous Plants of the United States and Canada, JOHN M. KINGSBURY
Principles of Biology, NEAL D. BUFFALOE
Radiotracer Methodology in Biological Science, C. H. WANG AND DAVID L. WILLIS
Readings in Ecology, EDWARD J. KORMONDY, ED.
Selected Botanical Papers, IRVING W. KNOBLOCH
Selected Papers on Virology, NICHOLAS HAHON
A Synthesis of Evolutionary Theory, HERBERT H. ROSS

Concepts of Modern Biology Series

*Behavioral Aspects of Ecology,** PETER H. KLOPFER
Molecular Biology: Genes and Chemical Control of Living Cells, J. M. BARRY
Processes of Organic Evolution, G. LEDYARD STEBBINS

Foundations of Modern Biology Series

Adaptation, 2nd ed., BRUCE WALLACE AND A. M. SRB
Animal Behavior, 2nd ed., VINCENT DETHIER AND ELIOT STELLAR
Animal Diversity, 2nd ed., EARL D. HANSON
Animal Physiology, 2nd ed., KNUT SCHMIDT-NEILSEN
The Cell, 2nd ed., CARL P. SWANSON
Cell Physiology and Biochemistry, 2nd ed., WILLIAM D. MCELROY
Chemical Background for the Biological Sciences, EMIL H. WHITE
Growth and Development, 2nd ed., MAURICE SUSSMAN
Heredity, 2nd ed., DAVID M. BONNER AND STANLEY E. MILLS
The Life of the Green Plant, 2nd ed., ARTHUR W. GALSTON
Man in Nature, 2nd ed., MARSTON BATES
The Plant Kingdom, 2nd ed., HAROLD C. BOLD

*These titles are also in the PRENTICE-HALL INTERNATIONAL SERIES IN BIOLOGICAL SCIENCE. Prentice-Hall, Inc.; Prentice-Hall International, United Kingdom and Eire; Prentice-Hall of Canada, Ltd., Canada.

CONCEPTS OF MODERN BIOLOGY SERIES

William D. McElroy and Carl P. Swanson, Editors

G. LEDYARD STEBBINS
Professor of Genetics
University of California, Davis

processes of organic evolution

PRENTICE-HALL, INC., Englewood Cliffs, New Jersey

PRENTICE-HALL INTERNATIONAL, INC., *London*
PRENTICE-HALL OF AUSTRALIA, PTY. LTD., *Sydney*
PRENTICE-HALL OF CANADA, LTD., *Toronto*
PRENTICE-HALL OF INDIA (PRIVATE) LTD., *New Delhi*
PRENTICE-HALL OF JAPAN, INC., *Tokyo*

Current printing (last digit):
10 9 8 7 6 5 4 3

Library of Congress Catalog Card Number 66-16917

Printed in the United States of America
C-72336

preface

For more than a century evolution has been both a cornerstone of biology and a focal point for conflicting theories. As we begin the second century following the *Origin of Species,* these theories have been sifted and combined to produce a synthesis upon which evolutionists in all disciplines of biology—taxonomy, genetics, cytology, ecology, and paleontology—are beginning to agree, at least in respect to its broad outlines. In the present volume this synthesis is offered to students who have had some training in biology and have mastered the basic principles and vocabulary of mendelian genetics. I owe a considerable debt to those who have been leaders in formulating it, particularly the late R. A. Fisher, Sir Julian Huxley, Theodosius Dobzhansky, Sewall Wright, Ernst Mayr, Edgar Anderson, and G. G. Simpson, both for the inspiration and guidance which reading their publications has given me, and for those personal discussions with most of them which have helped me immensely to formulate my own ideas. The present version of the synthetic theory is, however, strictly my own, and I take full responsibility for any ideas which may appear to deviate from those expressed in other publications.

I believe that we have entered a new age of evolutionary studies. The framework has been solidly built and is not likely to be either destroyed or radically modified by future research. Furthermore, exact scientific methods, based upon carefully made predictions and precise, quantitative experiments designed to test them, can now be used in many types of research on evolutionary problems. These methods need to be exploited to the full. In particular, they need to be aimed toward producing a new synthesis between, on the one hand, environmental biology and population genetics which are now the backbone of research in evolution and, on the other, comparative molecular biology which, in my opinion, is likely to become the nerve center of evolutionary theory in the future. I hope that this little volume will do two things. First, it may acquaint students and others who want to know about the "how" and "why" of evolution with the ideas which specialists in the field now hold, and with the kinds of facts upon which these ideas are based. Second, perhaps it can serve as a springboard from which adventurous biologists who are just now learning the basic framework of our science can in the future dive into the depths of the unknown and bring to the surface the treasures of new evolutionary facts and ideas which are hidden there.

I should like to acknowledge with particular thanks the help of my colleague Alex S. Fraser for his critical reading of the entire manuscript and for numerous helpful ideas which he has given me. I should like also to thank Robert W. Allard, Bernard John, Reuben Stirton, D. E. Savage, Sherwood Washburn, Paul Scotti, my wife Barbara, and members of the class in Evolution, Genetics 103, for the spring of 1965, who have read parts or all of the manuscript and have contributed many helpful suggestions. I owe thanks also to Miss Sally Broadbent for her careful and faithful typing of the manuscript.

G. LEDYARD STEBBINS

contents

processes of organic evolution

The synthetic theory of evolution and its development

CHAPTER 1

Modern biology has two unifying concepts. One is the concept of organization. This tells us that at every level, from the molecule through the supra-molecular organelle, the cell, the tissue, the organism, the individual, and up to the population or the society, the properties of life depend only to a small degree upon the substances of which living matter is composed. To a much greater degree living things owe their nature to the way in which the components are organized into orderly patterns, which are far more permanent than the substances themselves. The other unifying concept of biology is that of the continuity of life through heredity and evolution. This tells us that organisms resemble each other because they have received from some common ancestor hereditary elements, chiefly the chromosomes of their nuclei, which are alike both in respect to the substances which they contain and the way in which these substances are organized. When related kinds of organisms differ from each other, this means that in the separate lines of descent from their common ancestor changes in the hereditary elements have taken place, and these changes have become established in whole populations.

THE FACT OF EVOLUTION At the outset we should realize that the great majority of biologists accept as demonstrated the fact that organisms have evolved. To be sure, no biologist has actually seen the origin by evolution of a major group of organisms. Nevertheless, races and species have been produced by duplicating in the laboratory and garden some of the evolutionary processes known to take place in nature. The reason that major steps in evolution have never been observed is that they require millions of years to be completed. The evolutionary processes which gave rise to major groups of organisms, such as genera and families, took place in the remote past, long before there were people to observe them. Nevertheless, the facts which we know about these origins, some of which will be discussed in Chapter 7, provide very strong circumstantial evidence to indicate that the processes which brought them about were very similar to those found in modern groups of animals and plants which are evolving all around us today.

The state of our knowledge of the major steps of evolution which took place in the past is much like that of our knowledge of ancient history. We do not have any reliable eye-witness accounts of the events which produced the rise and fall of the civilizations of ancient Egypt, Sumeria, Babylon, and Crete, and the contemporary written records of these ancient times are fragmentary and unreliable. Nevertheless, the indirect evidence that these civilizations existed is so strong that it is accepted by scholars and laymen alike. Moreover, we teach as historical fact many of the probable events connected with their rise and fall. The evidence which biologists now have about the rise and extinction of major groups of animals in past geological eras is of a very similar nature, and carries with it about the same degree of high probability.

Since Darwin developed his theory of evolution more than a century ago, biologists have studied this subject in two different ways. Some have been interested in the course of evolution. By comparing a multitude of different kinds of animals or plants, they have tried to work out the evolutionary family tree of some particular group, such as the horses, the cone bearing trees, or mankind. Other biologists have asked themselves the question: "What makes evolution go?" By means of observations and experiments of various sorts on populations of living organisms, they are learning about the processes of evolution and the mechanisms responsible for them. The present book will consider only this second approach to the study of evolution. The study of the course of evolution, or phylogeny as it is often called, must be carried out in a somewhat different way in each separate group of organisms. Such studies have yielded few general principles which apply equally well to all kinds of living things. On the other hand the processes of evolution are in many ways similar even when we compare such different forms as bacteria, flies, grasses, and mammals. Consequently, an understanding of these processes is of far greater importance to the general biologist than is knowledge of the particular ancestry of any group.

THE MODERN SYNTHETIC THEORY OF EVOLUTION

The modern, synthetic theory of evolution recognizes five basic types of processes: GENE MUTATION, CHANGES IN CHROMOSOME STRUCTURE AND NUMBER, GENETIC RECOMBINATION, NATURAL SELECTION, AND REPRODUCTIVE ISOLATION. The first three provide the genetic variability without which change cannot take place; natural selection and reproductive isolation guide populations of organisms into adaptive channels (Figure 1-1). In addition, three accessory processes affect the working of these five basic processes. MIGRATION of individuals from one population to another, as well as HYBRIDIZATION between races or closely related species both increase the amount of genetic variability available to a population. The effects of CHANCE, acting on small populations, may alter the way in which natural selection guides the course of evolution. The purpose of the present book is to review our knowledge of each of these processes, and to show how they are interrelated with each other. The more we know about the five basic processes, the less reason we have for believing that any other basic processes remain to be discovered. We do not need to search any more for hidden causes of evolution. Nevertheless, we do need to understand much more about the way in which known processes interact with each other.

At the outset we must recognize that at least in higher organisms, and perhaps in microorganisms as well, the three processes, mutation, genetic recombination, and natural selection, are equally indispensable for evolutionary change to take place. Speculations as to which of the three is the most important are completely pointless. Their interrelationships can be well

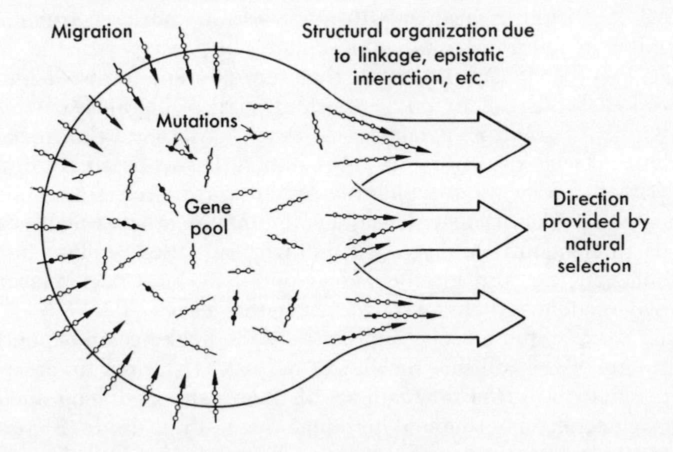

Fig. 1-1. Diagram to illustrate how the four basic processes—(1) mutation; (2) genetic recombination, which results from intercrossing between individuals of the population as well as between them and occasional new genotypes which enter by migration; (3) Structural changes in the chromosomes, which affect linkage and epistatic interaction of genes; and (4) Natural selection—interact to produce a progressive change in the population which keeps it adapted to the changing environment.

illustrated by the following analogy. An evolutionary line of organisms which is changing through eons of time can be likened to an automobile being driven along the highway. Mutation then corresponds to the gasoline in the tank. Since it is the only possible source of new genetic variation, it is essential for continued progress, but it is not the immediate source of motive power. This source is genetic recombination, through the shuffling of genes and chromosomes which goes on during the sexual cycle. Since this process provides the immediate source of variability upon which selection exerts its primary action it can be compared to the engine of the automobile. Natural selection, which directs genetic variability toward adaptation to the environment, can be compared to the driver of the vehicle. Various lines of evidence, which will be explained in a later chapter, indicate that structural changes in the chromosomes, which change the order of arrangement of the genes along them, can have profound effects upon the interrelationships between genetic recombination and natural selection, and so can be compared to the transmission and accelerator of the automobile. Finally, reproductive isolation, which includes all of the barriers to gene exchange between populations, has a canalizing effect similar to that which the highway, with its limits and directive signs, exerts upon the driver of the automobile, thus permitting several vehicles to drive in the same direction at the same time.

CHARLES DARWIN, THE FOUNDER OF EVOLUTIONARY THEORY

The synthetic theory of evolutionary dynamics just outlined was not thought up by any one scientist. Rather, it has itself evolved during the last century through an accumulation of factual evidence and theoretical conclusions. We can begin its history with the work of Charles Darwin. To be sure, a number of naturalists and philosophers, beginning with the ancient Greeks, had thought of the possibility that the different forms of life have evolved from each other. The most important of these before Darwin's time was Jean Baptiste Lamarck, a renowned French naturalist, who in his later years produced a well-known theory of evolution. It was based largely upon the concept that modifications which the organism acquires in adaptation to the environments which it meets during its lifetime are automatically handed down to its descendants, and so become part of heredity. For instance, Lamarck believed that the giraffe had acquired its long neck because its ancestors had stretched their necks to reach the leaves of tall trees. As a result of this exercise, the muscles and bones of the neck became abnormally developed. The effects of this development were believed to have been transmitted somehow to the descendants of these neck-stretching ancestors, and so to have become a permanent hereditary trait of the giraffe (Figure 1-2). In a similar way, a Lamarckian theorist would explain the dark skin color of some races of man by assuming that the ancestors of these races had been exposed repeatedly to the strong rays of the tropical sun, thus acquiring a tan which was transmitted to their descendants. This concept of the inheritance of acquired modifications, though not completely disproved, has so little evidence in favor of it that few biologists of the present day regard it as an important factor in evolution. Although Darwin and his contemporaries did not have enough knowledge to reach this conclusion, most of them paid little attention to Lamarck's theory because it consisted entirely of armchair speculation, and Lamarck produced no factual evidence to support it.

Charles Darwin's approach to the theory of evolution was quite different from that of Lamarck. His life was devoted to a continuous series of observations, experiments, and accumulation of facts by correspondence with other naturalists. All of his efforts were directed toward a better understanding of the adaptation of organisms and their evolution. He began the observations which led to his theory of evolution at the age of 22, when he was naturalist on *H.M.S. Beagle*, a ship which the British navy sent on a five year cruise around the world. The book which Darwin wrote about this voyage, now a classic, is fascinating reading for anyone interested in natural history. At the same time it gives us valuable insight into the kinds of observations which led Darwin to his theory. During an excursion into the Argentine pampas from the city of Buenos Aires he became greatly impressed with the revolutionary change in the country's vegetation which had been produced by the arrival of civilized man from Europe. Immigrant species of plants brought in by accident from their native European home were overrunning the fields and roadsides driving out the native South American species. Later, the *Beagle* sailed down the barren shores of Patagonia, near the southern tip of South America, where Darwin saw some of the richest fossil beds in the world. The bones of extinct species of mammals were lying exposed on the cliffs by the thousands, giving ample evidence that the animals of past ages were different from those of present-day South America but clearly related to them.

After passing through the Straits of Magellan the *Beagle* sailed northward along the Pacific coast of South America, and spent five weeks at the Galapagos Islands, 600 miles west of the coast of Ecuador. Here Darwin became aware of the fact that the principal animals were different from those of the South American coast. Huge lizards were everywhere, some of them living on the actual seashore and feeding on marine life. These lizards are peculiar

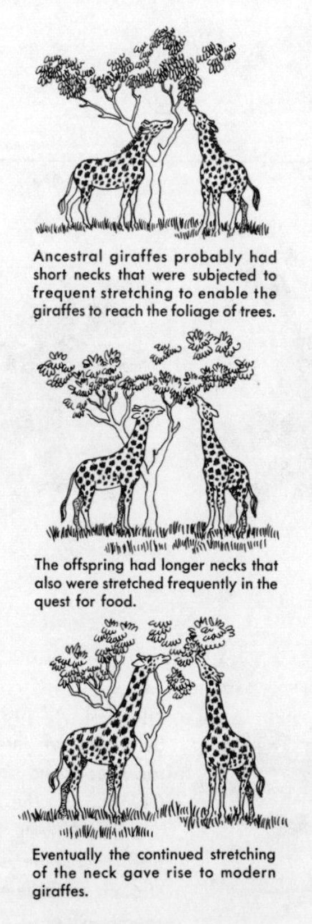

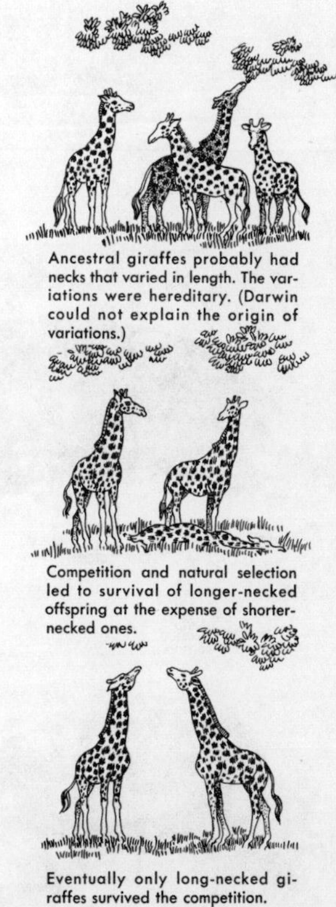

Fig. 1-2. Illustration showing how the origin of the giraffe's long neck is explained according to the now discredited Lamarckian theory of the inheritance of acquired modifications and according to the generally accepted Darwinian theory of natural selection. From *Biological Science: An Inquiry into Life* (Harcourt Brace and World).

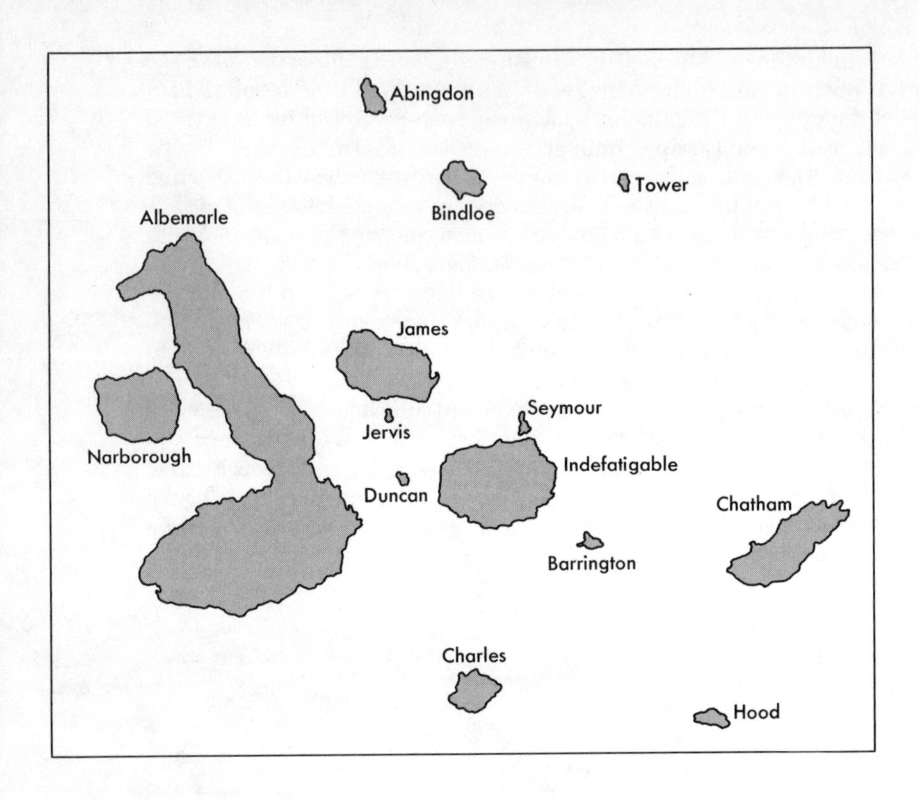

Abingdon

Duncan

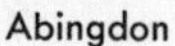

Albemarle

Fig. 1-3. Drawings showing the characteristics of three different species of tortoises found on different islands in the Galapagos. The longer necked species live in relatively dry places and feed on tree cacti; the species with the short, straight neck lives in moister regions and feeds on dense, low growing vegetation.

to the Galapagos, but their nearest relatives are found in South America. The giant tortoises for which the islands are named were the commonest animals of the interior of the islands. Not only do these belong to a completely different group from any inhabiting the American mainland, but in addition each separate island of the Galapagos archipelago has its own particular race of tortoise (Figure 1-3). Darwin observed that these races of tortoise are so different from each other that even the sailors who often visited the islands could tell by its appearance from which island a particular tortoise came. The plant life is distinguished by the fact that the largest and commonest trees of the interior forests belong to a distinctive genus (*Scalesia*) of the sunflower family (Compositae). In other parts of the world this family produces only herbs and small shrubs. Like the tortoises, these tree sunflowers include a separate race or species for almost every one of the islands. In the book which he wrote about his journey Darwin comments on these facts as follows:

> Why, on these small points of land, which within a late geological period must have been covered by the ocean, which are formed of basaltic lava, and therefore differ in geological character from the American continent, and are placed under a peculiar climate,—why were their aboriginal inhabitants, associated, I may add, in different proportions both in kind and number from those on the continent, and therefore acting on each other in a different manner—why were they created on American types of organization?

And later:

> Reviewing the facts here given, one is astonished at the amount of creative force, if such an expression may be used, displayed on these small, barren, and rocky islands; and still more so, at its diverse yet analogous action on points so near each other.

From these comments we can see that in 1837, when he was 28 years old, Darwin was already questioning the then prevalent doctrine of special creation. A year later (1838) he read a book on human populations by the English clergyman, Thomas Malthus, in which the author pointed out that unless checked by disease, war, famine, or conscious control of reproduction, the number of people on the earth would in a short time increase so much that there would be "standing room only." Darwin could easily obtain figures to show that the same would be true of any kind of animal or plant, even for such slowly reproducing species as the elephant. As a naturalist, he was well aware of the fact that organisms are adapted to their environment, and that this adaptation often takes the form of elaborate and bizarre structures and activities. By careful observation of many species in nature, in his garden and as animals kept in cages about his home, he convinced himself that the individuals of any population differ slightly from each other in many characteristics, including those which contribute toward adaptation. He made the logical deduction, therefore, that the factors which check the increase of numbers in a species act more strongly on those individuals which are relatively poorly adapted and favor those which are best fitted to their environment. Since these favored individuals will leave more offspring than their less well adapted associates, this process of natural selection, continued over many generations, should evolve ever more perfect and complex adaptations, and so bring about progressive evolution.

Many scientists, as soon as such a fine idea occurred to them, would immediately rush to print with it. But this was not Darwin's way of working. He started his first notebook on the transmutation of species in 1837, and in 1844 wrote his first essay on the subject. He never published this essay because he felt that he did not have enough evidence for his theory. Instead, he devoted his entire attention to accumulating systematically the necessary evidence. Having inherited a fortune, he could live in his country home and devote his entire time to gathering all kinds of information which might bear upon his theory. He realized that natural selection, as he conceived it, is very much like artificial selection, by means of which animal and plant breeders had been able to produce much altered and improved breeds of domestic animals and cultivated plants. He therefore became a pigeon breeder and studied carefully both the way in which the fancy breeds of his day had been produced, and the anatomical characteristics which distinguished them from each other and from wild pigeons. He chose the pigeon because all domestic varieties of this bird are certainly derived from a well-known wild species of Europe. Darwin was astonished to discover that in certain anatomical features the breeds of pigeons differ from each other to a greater degree than do many species and even families of wild birds. This convinced him that selection, either artificial or natural, if continued long enough and with great enough intensity, could bring about the kinds of differences by which naturalists are accustomed to recognize various wild species.

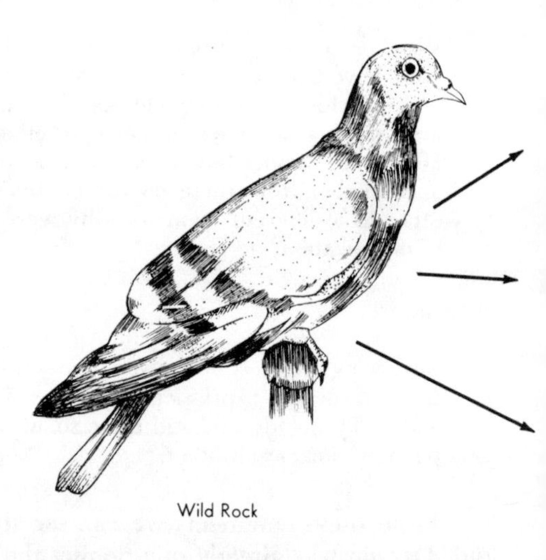

Wild Rock

At the same time, Darwin was gathering facts of many other kinds which might bear on his theory. Through his own observations of fossil animals, as well as by studying the publications of contemporary zoologists and paleontologists, he learned much about prehistoric animals and their relationships with modern species. Sir Joseph Hooker, an equally eminent botanist, supplied him with valuable evidence from the plant kingdom.

Twenty years after his discovery of Malthus' book Darwin was still gathering facts for a large and complete treatise of his own when he received a manuscript from Alfred Russell Wallace, another British naturalist who was at that time studying the fauna of the East Indies. In this manuscript, Wallace set forth clearly the same theory as Darwin's, which he had conceived independently; that species of animals have evolved from each other through

Fig. 1-4. Variation among the breeds of domestic pigeon, a subject of investigation by Darwin which confirmed for him the strong influence on variation which can be exerted by natural selection. From Wallace and Srb, *Adaptation* (Prentice-Hall, Inc.).

Fig. 1-5. Photograph of Charles Darwin in middle age.

the action of natural selection. Rather than competing with each other to be the first to publish this theory, Darwin and Wallace agreed that their papers should be read jointly at an historic meeting of the Linnean Society of London, on July 1, 1858. A year later (1859), Darwin published his famous classic, *The Origin of Species,* which, though much shorter than he had originally intended it to be, nevertheless contained a great wealth of factual material.

Both biologists and other people reacted immediately and violently to this book. Some found in it answers to questions which they had long been asking themselves; others defended tenaciously their previous concepts about the immutability of species. The criticism from the church, on the grounds that Darwin's theory contradicts the story of creation told in the Book of Genesis, led to violent polemics in Darwin's day and have continued into the present century.

Darwin took most seriously those criticisms which were levelled at his theory of natural selection. In successive editions of *The Origin of Species* he spent many words answering these critics. We usually read the sixth edition, published in 1872, which contains all of these answers. At the same time, Darwin was continuing his research in a variety of directions. The natural history of earthworms, the forms of flowers in relation to their methods of pollination, the effects of cross- and self-fertilization in plants, the differences between males and females of the same species; all these he studied carefully in order to find an explanation of their origin based upon natural selection. Reviewing the life of Darwin, we cannot fail to admire the breadth of his knowledge, the scope of his investigations, and the persistence with which he directed these toward a single goal. Darwin succeeded in convincing the world of the existence and great significance of organic evolution because of his profound knowledge of natural history; his ability to accumulate a wide variety of facts by means of observation, experiment, and correspondence with other naturalists; and particularly his ability to synthesize these facts into a coherent whole. Another important factor was that the naturalists of the mid-nineteenth century, who were finishing the most intensive series of explorations of the world's flora and fauna which has ever been made, were in the right mood for seeking and accepting explanations for the tremendous diversity of living things which was now laid before them.

THE IMPACT OF MENDELISM

Nevertheless, Darwin's theory had one serious flaw. He knew nothing about the causes of hereditary variation, and his opinions on this subject were neither logical nor consistent. At times he accepted, and in other writings he rejected Lamarck's notion of the inheritance of adaptations acquired during the lifetime of the individual. Along with most practical animal breeders and students of human heredity of his day, he regarded the hereditary substances as fluid in nature, and the observed intermediate nature of hybrids between races or breeds as resulting from a mixture of parental fluids in their bodies. One of Darwin's severest critics, Fleeming Jenkin, pointed out that selection could not sort out superior fluids from a mixture, and so, according to prevailing theory, could not be the means by which superior types could be derived from crosses between breeds. Because he knew so little about heredity, Darwin could not provide a satisfactory answer to this criticism.

The answer, that heredity is determined not by fluids but by particulate genes, already existed in the garden of an obscure monastery in Brunn, Czechoslovakia, in the experiments of Gregor Mendel. But Darwin knew nothing about this.

Furthermore, the rediscovery of Mendel's laws of heredity by Correns, de Vries, and Tschermak in 1900 did not immediately reinforce Darwin's theory, but on the contrary, placed it in a temporary eclipse. This was because both Mendel and the rediscoverers were studying characteristics which, as differences between natural populations, are relatively uncommon. Mendel crossed a tall with a dwarf strain of peas and found in the second generation a ratio of three tall to one dwarf plant. He could, therefore, conclude that the parents differed in respect to a single gene controlling size. When we cross a tall human being with a short one, or a tall race of wild yarrow with a dwarf alpine plant of the same species, we find that no simple Mendelian ratio can be found in the second hybrid generation. The difference between tallness and shortness in these species is governed by many genes, each one of which has only a slight effect on size. Darwin's theory was based upon his sound observation that most differences between natural populations are quantitative in nature, but neither he nor the earliest of the Mendelian geneticists understood why this is so.

Two other events in the early history of genetics combined to reinforce the skepticism of these geneticists toward Darwin's theory. The first of these was the discovery by de Vries that in the evening primrose (*Oenothera*) new types, differing markedly from their parents in a number of characteristics, can arise at a single step. He called these changes mutations, a term which has ever since been used chiefly for changes in single genes. We now know that many mutations occur which are not abrupt alterations of one or many characteristics, but produce slight, barely perceptible modifications of the visible structures. In 1905, a few years after the work of de Vries, the Danish botanist, Johannsen, tried applying Darwin's theory to altering the genetic nature of the garden bean. He selected the largest and the smallest beans from the seed lot of a single variety, and grew offspring from them. In the first generation, the larger beans produced slightly larger offspring than the smaller ones, but in later generations selection for large and small size had no effect on the offspring. From this experiment, Johannsen and other early geneticists concluded that the fluctuating variation which Darwin had observed in natural populations, and which he considered to be the basis of natural selection, was actually not hereditary at all, but merely due to the effect of the environment on individual organisms. Putting together the discoveries of de Vries and Johannsen, they argued that evolution takes place through the spontaneous origin of new types, which differ radically from their parents in several characteristics. Natural selection, according to them, has merely the negative function of eliminating those types which are unfit to survive.

This hypothesis, which was known during the first quarter of the twentieth century as the mutation theory, has not withstood the test of time for a number of reasons. The mutations which de Vries found in the evening primrose were later found to be not the result of new genetic variation at all, but merely a peculiar type of genetic segregation due to the fact that this plant has a very anomalous type of chromosome behavior. Later on, Morgan and his associates found true mutations in the fly, *Drosophila*, which are actually spontaneous alterations of genes, and they have since been found in a large number of different organisms. But nearly all mutations which produce large, conspicuous changes also make the organisms bearing them weaker and unable to compete with their unchanged associates. Moreover, geneticists have recently carried out experiments on selection for size and other characteristics with mice, flies, corn, and other organisms, and have obtained entirely different results from those which Johannsen got in his experiments with beans. Selection in these organisms has produced progressive changes over as many as fifty to a hundred generations. This difference is explained by the fact that the garden bean is self fertilizing and has been selected by breeders over many generations for uniformity and constancy. In this way, differences due to genes with small effects have been artificially eliminated.

In natural populations, most of which are cross fertilizing and have not been subjected to such rigid selection, a large "gene pool" of genetically controlled variation exists at all times. Most of the adaptively important mutations do not produce entirely new types, but merely add quantitatively to the gene pool for already existing variations. Natural selection has the positive, creative function of sorting out a few adaptive gene combinations from the infinite number of possibilities inherent in the gene pool.

The numerous facts which are the basis of the statements just made were acquired by several geneticists during the entire first half of this century. H. Nilsson-Ehle in Sweden and E. M. East at Harvard showed that quantitative hereditary variation is due to the simultaneous action of many separate genes, each of which has a small effect by itself. J. B. S. Haldane and R. A. Fisher in England, Sewall Wright in the United States, and S. S. Chetverikov in the Soviet Union all showed by mathematical calculations that evolutionary change must depend not only on the origin of genes with new effects, but also on changes in the frequency of all of the genes in the population. Such changes cannot be brought about by mutation alone, with selection acting only to eliminate unfit mutants, but must depend upon the continuous action of natural selection, acting at measurable intensities. Finally, various geneticists developed techniques for raising large populations of drosophila flies of known genetic composition in cages under carefully controlled environments. In this way, a tremendous wealth of experimental evidence has been amassed in favor of the theories of population dynamics and evolution worked out mathematically by Haldane, Fisher, Wright, and Chetverikov. Thus the modern, synthetic theory of evolution has acquired a solid basis of scientific fact.

THE CONFLICT BETWEEN DARWINISM AND MENDELISM, AND ITS RESOLUTION

One of the most curious phases in the history of modern biology is the violent conflict concerning the nature of hereditary variation and the processes of evolution which took place between the early Mendelians, chiefly de Vries, Bateson, and Johannsen, and the contemporary Darwinian naturalists, such as David Starr Jordan and H. F. Osborn, as well as statistical biologists such as Francis Galton and Karl Pearson. Its cause was two-fold. In the first place, the members of the two opposing camps were biologists with opposing temperaments and philosophical attitudes. De Vries, Bateson, and Johannsen were basically experimentalists, to whom precision in experimental design and care in interpretation of results were of prime importance. Because their field was young, they could maintain these standards only at the expense of breadth, so that they had to base generalizations on a few examples. The naturalists were already aware of a fact which these experimentalists did not realize, and which became obvious only during the subsequent twenty years. This fact is that the examples of simple genetic segregation and constancy which Mendel, de Vries, and Johannsen found in their experiments were not representative of hereditary variation as it exists in natural populations of cross breeding organisms. On the other hand, the naturalists who promoted Darwin's theory in the early part of this century were describers and cataloguers who had no conception at all of the precision of thinking which is required to design a good experiment and to interpret its results correctly. Consequently, they failed to appreciate the significance of either Mendelian genetics or the mutation theory, or to understand how these concepts could be modified to interpret the pattern of variation in nature, which they knew very well.

The conflict was caused also by the different outlook of the two groups on variation itself. The Mendelians were looking upward from the genes, the naturalists downward from the phenotypes, and neither had the least conception of the enormous complexity of biological processes which separates the two in higher animals and plants. This complexity is illustrated in the following diagrams. The first of them (Figure 1-6) illustrates two empirically demonstrated facts and their probable explanation. One of these which has already been mentioned, is that most visible differences between populations, races, and species of higher animals and plants are governed by many different genes, each of which contributes to a different one of the numerous metabolic processes which can affect such characteristics as body size, sexual maturity, and intensity of skin pigmentation. In Figure 1-6, this fact is represented by the presence of several arrows pointing toward each one of the characters represented at the right hand side of the diagram. Conversely many differences controlled by single genes affect a large number of different characteristics of the adult organism. This is because a disturbance or a change in the rate of some basic metabolic or synthetic process, such as the formation of cartilage or bone in a higher animal or of lignified cell walls in a higher plant, is bound to affect many different organs in various ways. In many instances, particularly in the case of the synthesis of enzymes and other basic processes of cellular metabolism, a gene-controlled difference can have a "feedback" effect on the rate of duplication or activity of the genes themselves. Such feedbacks are represented in Figure 1-6 by arrows pointing from right to left. Figure 1-7 supplements Figure 1-6 with a representation of specific metabolic processes which must take place in a higher organism between the initial activity of genes and the final expression of morphological characteristics. No effort has been made to exaggerate complexity in making this diagram; if anything it is an oversimplification of the network of processes which actually takes place.

This developmental complexity is important to students of evolution for the following reason. Natural selection acts at all stages of development, but much of its action is on the final characteristics of the adult organism. Evolutionary change, however, depends on alterations of the frequency of particulate genes in the population. The complexity of the pattern of developmental processes illustrated in Figures 1-6 and 1-7 is a measure of the degree of indirectness of the effects of natural selection on evolution. Because a single visible difference is usually controlled by many genes, any action of selection which affects this difference will automatically alter the frequency of several different genes in the population. Conversely, if the frequency of a gene is altered by the selective advantage or disadvantage of one of its end effects, several other characteristics over which it also exerts partial control will be simultaneously altered. Before biologists can understand fully the ways in which selection can produce visible differences between races and species, they must explore more deeply into a great *terra incognita* of modern biology; the complex network of pathways between genes and characters.

The resolution of these complex difficulties, and the synthesis of a coherent theory of evolution which takes into account all of the pertinent facts of modern biology, has been the work of several biologists during the past thirty years. The first edition of Dobzhansky's now classic book, *Genetics and the Origin of Species,* which appeared in 1937, set the stage, and stimulated biologists in several fields to contribute to the synthesis. Books by

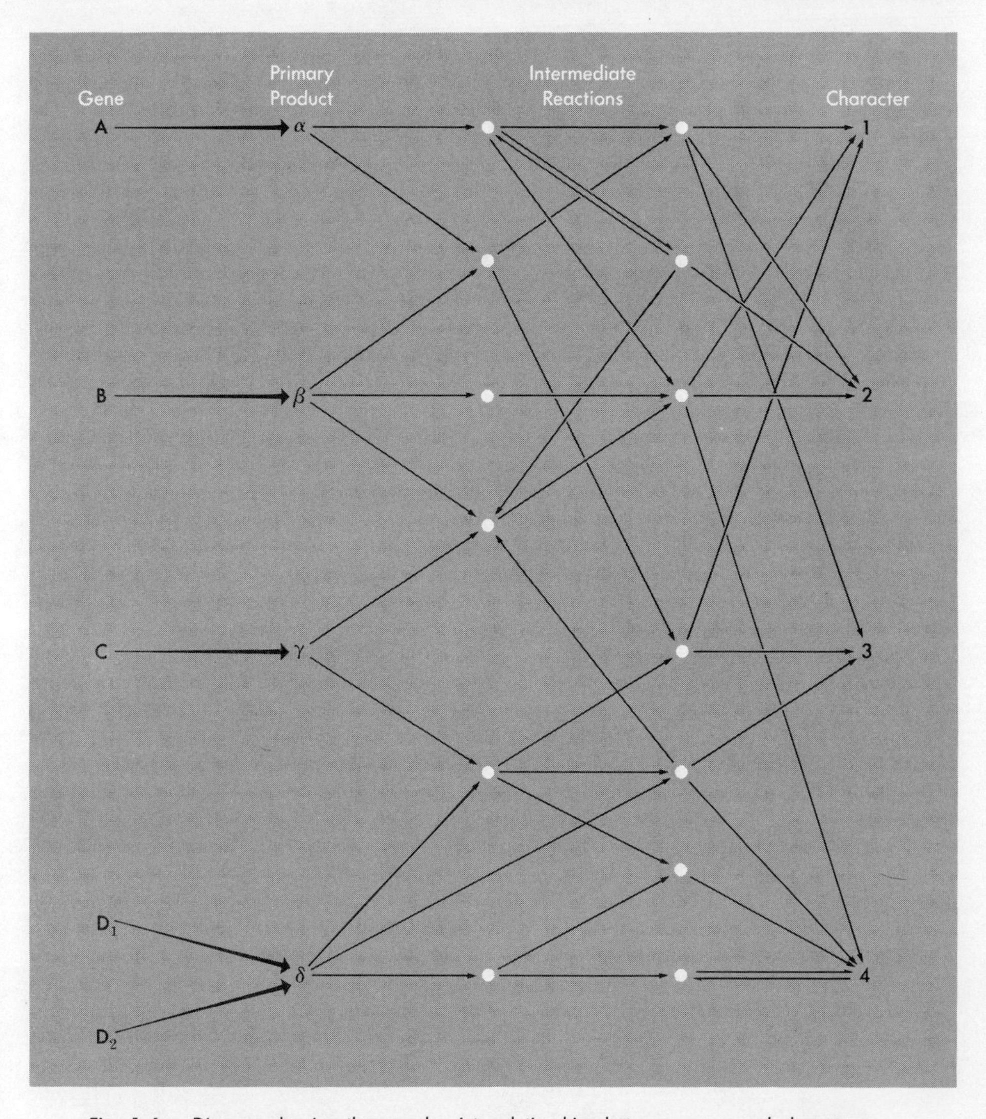

Fig. 1-6. Diagram showing the complex interrelationships between genes and characters which result from the existence of genes with multiple or pleiotropic end effects, from the multiple factor basis of inheritance of most adult characteristics, and from the frequent presence of "feed back" interactions by which intermediate products of the action of one gene can affect primary products of another gene.

zoologists Julian Huxley and Ernst Mayr showed how the modern theory could explain the origin of variation patterns in higher animals, and the present writer attempted to do the same for higher plants. A leading paleontologist, George Simpson, showed in two books that the fossil record of higher animals is best explained by assuming that throughout the evolutionary history of living things those same processes took place which were

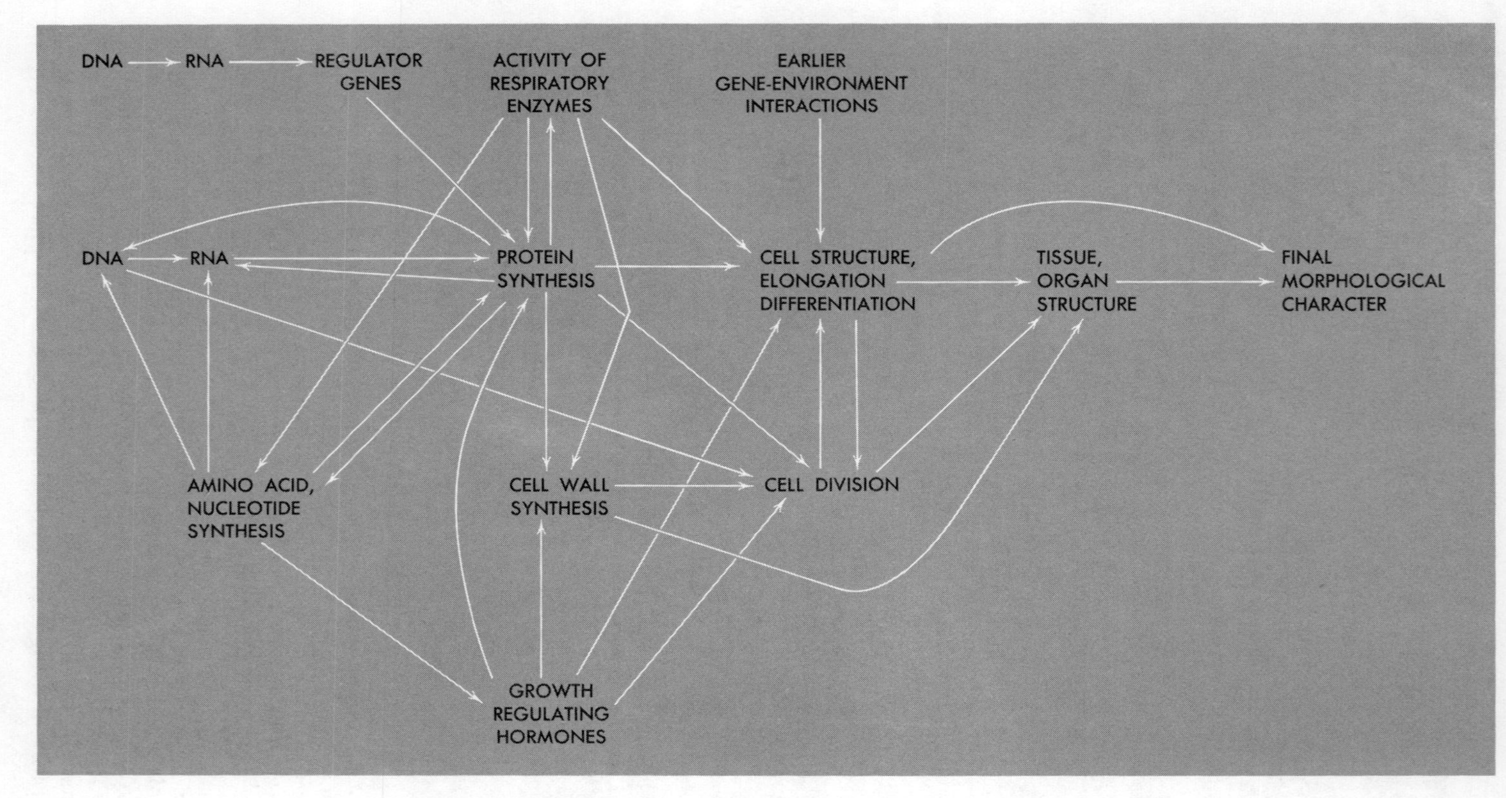

Fig. 1-7. Diagram showing some of the metabolic processes which intervene between the primary action of genes and the final expression of characters in a higher organism.

being experimentally demonstrated by many workers in populations of contemporary animals. In the last few years, Bernhard Rensch has made a strong case for the belief that the evolution of genera, families, and higher categories of animals has taken place through an extension into long periods of time of those same processes which at any one time level govern the origin of races and species.

Chapter Summary

Modern biologists accept evolution as a fact, as well demonstrated as the facts of ancient history. The basic processes of evolution are five; (1) mutation and (2) genetic recombination which are the sources of variability, but do not provide direction. They contribute variability to a gene pool represented by the variant individuals composing any cross fertilizing population in nature. (3) Chromosomal organization and its variation, which affect genetic linkage, produce orderly arrangements of variation in the gene pool, which changes its composition through the guidance of (4) natural selection. Limits to the direction in which selection can guide the population are set by (5) reproductive isolation.

Charles Darwin in his book, *The Origin of Species,* provided evidence which convinced most biologists that evolution has occurred and produced the theory of natural selection to explain it. The weakness in his theory, ignorance of the nature of heredity, was removed when, in the 1920's and 1930's, Mendelian principles of inheritance were correctly applied to populations and used to explain genetic variability in nature. A conflict which existed in the first quarter of the twentieth century between Darwinian naturalists and early Mendelian geneticists was resolved by this research in population genetics. The modern synthetic theory of evolutionary processes, which is the theme of this book, is the result.

Questions for Thought and Discussion

1. What characteristics of Darwin's work and of the period in which he lived contributed the most to the favorable reception which his theory received, as compared to Lamarck and others of Darwin's predecessors?

2. Do you think that Darwin would have radically changed the last edition of *The Origin of Species* (published in 1872) if he had known about Mendel's discoveries when they were made (1865)? Give reasons for your answer.

3. What are the similarities and differences between Darwinian and modern concepts of evolutionary processes? In what way has Mendelian genetics contributed toward establishing these differences?

CHAPTER 2

The sources of variability

If we look at any population of cross fertilizing, sexually reproducing organisms, we are impressed with the differences among their individuals. Among the trees of any forest, the cattle in a pasture, a group of people in a classroom, theatre, or city street, no two look alike. Geneticists are well aware of the fact that these differences are produced by two very different kinds of factors. To some extent they are due to differences in nutrition, crowding, disease, and other factors of the environment. They are also based partly upon differences in genetic makeup. Whenever an outbreeding population is subjected to genetic analysis, numerous differences in genes are found among its individuals. In genetic terms, two or more alleles exist at a large proportion of loci among the individuals of any outbreeding population. The sum total of different alleles is called the GENE POOL of the population.

A third reason for variability in populations is that genes are not completely stable particles. They may occasionally change or mutate to a different state, or allele, which may be either new or already present in the population. Such MUTATIONS occur so rarely that their contribution to the variability of a population in a single generation is negligible. Nevertheless, mutations are the only source of new genetic variability, and without them evolution could not progress for very long. Consequently, they must be considered in any discussion of the processes of evolution. The evolutionist regards variability in populations as possessing three components: (1) Environmental, due to PHENOTYPIC PLASTICITY, (2) Genetic and Recombinational, (3) Genetic and Mutational. The first two contribute about equally to the variation seen in most crossbreeding populations. By comparison, the immediate contribution of mutation, i.e. its contribution to the total amount of phenotypic variation per generation, is minute. The present chapter will discuss environmental and mutational variability; the more complex topic of recombinational variability will be taken up in the next chapter.

THE NATURE AND SIGNIFICANCE OF ENVIRONMENTAL MODIFICATION

One of the most important facts about the phenotypic variation caused by the environment is that it is often adaptive. It helps the organism to adjust better to the particular environment in which it lives. We are all familiar with the fact that exercise stimulates the development of muscles and reflex reactions, thus enabling a person to perform better the movements which he has practiced many times. Some species of plants can vary in a striking fashion by altering their patterns of growth in response to changes in their environment. For instance, certain species of buttercup which live in shallow pools or on the edges of quiet ponds, where the water level is subject to great changes, produce finely dissected leaves when their leaves are deeply submersed, and entire leaves when they are near the surface of the water (Figure 2-1). If these species are grown under controlled conditions, the experimenter can change the shape of the leaves at will by altering the environment. The adaptive significance of these changes is that they maintain an optimum balance between the two factors most needed for photosynthesis, light and water. For the submersed leaf, only light is a limiting factor. Hence the larger is the surface of the leaf relative to its volume, the larger is the proportion of cells and chloroplasts which are directly exposed to the relatively weak, filtered light which penetrates the water. On the other hand,

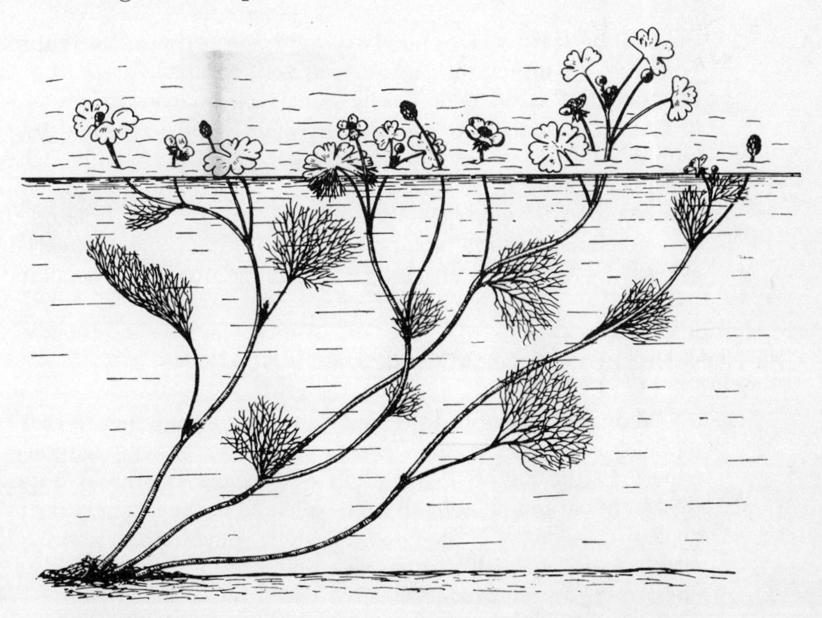

Fig. 2-1. Water buttercup (*Ranunculus aquatilis*) showing extreme difference in leaf shape between floating and submersed leaves. From Weaver and Clements, *Plant Ecology* (McGraw-Hill Book Company).

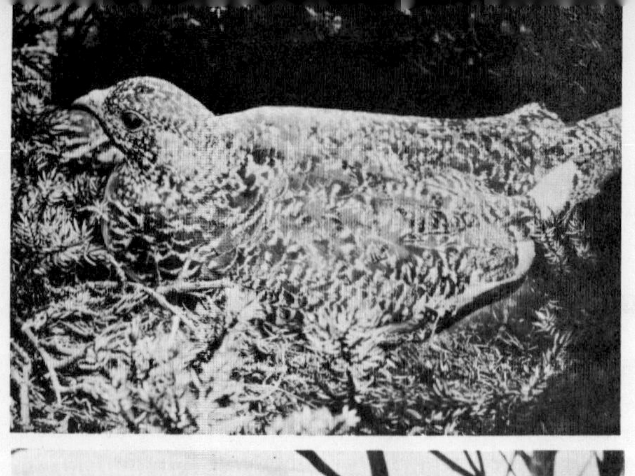

Fig. 2-2. Summer and winter plumage of ptarmigan. From Murphy and Amadon, *Land Birds of America* (McGraw-Hill Book Company).

the floating leaf receives plenty of light, but can lose water through evaporation from its upper surface. A relatively compact shape has the advantage of reducing this water loss. The same fact is illustrated by changes in color of certain animals. The arctic ptarmigan (Figure 2-2) and the varying hare both live in climates characterized by long, snowy winters, when the ground is white for long periods of time. They can respond to fall temperatures and light regimes by changing their coat color to white. Related species, which live in climates with much shorter periods of snow covered ground, are genetically incapable of making this change under any conditions.

THE NON-INHERITANCE OF ACQUIRED MODIFICATIONS

Many evolutionists, following Lamarck, have believed that such adaptive modifications can become incorporated into genetic heredity and so contribute to the adaptive evolution of species. The great improbability that this could happen in animals was explained in the last century by the German zoologist August Weismann, who pointed out that the germ cells of animals, which will eventually form eggs and sperm, become separated from the embryonic tissues which will give rise to other parts of the body at a very early stage in development. It is, therefore, very difficult to conceive of any way in which alteration of such bodily tissues as muscles could induce corresponding alterations in the gametes, and so be passed on to future generations.

The fact that repeated attempts to perform experiments which might demonstrate the inheritance of adaptive modifications in animals have never produced successful and repeatable results is in accord with Weismann's theory.

In higher plants, and particularly in such microorganisms as bacteria, separation between body and germ cells either does not occur at all or is confined to late stages of development. Consequently, Weismann's theoretical argument cannot be applied to these organisms. For many years, bacteriologists assumed as a matter of course that adaptive modifications in their cultures came about through the direct action of chemical substances in the medium. Now, however, many experiments have been carried out which have shown that such modification is usually brought about by the action of natural selection on spontaneous mutations. Some of these experiments will be described in Chapter 4 of this book. They show us that even in those organisms in which we would most expect to find it, the inheritance of acquired adaptive modifications is not a significant factor in evolution.

In higher plants the inheritance of acquired adaptations is not completely excluded as a factor of evolution. Many experiments have been performed in attempts to demonstrate the existence of this factor; recently a large proportion of them have been in Russia and other communist countries, under the influence of the Russian biologist T. D. Lysenko. Attempts to repeat such experiments in western countries have consistently failed, and eye witness accounts from western observers cast grave doubts on the validity of the experimental methods used by the Lysenko school. Recently, Lysenko has also become discredited in the Soviet Union. There are, however, other experiments, particularly one carried on at Aberystwyth, Wales, on the response of flax to fertilizers, in which the technical methods were beyond reproach. In these experiments some of the increased vigor which the plants acquired following heavy fertilization of the soil was transmitted to their offspring. The experiments with bacteria which are described in Chapter 4 suggest that changes in nuclear heredity are brought about much more easily by spontaneous or chemically induced genetic mutation than by any of the environmental influences which an organism normally faces during its lifetime. There is, however, some indication that quantitative characteristics can in some circumstances be influenced by hereditary factors in the cytoplasm. Consequently, environmental alterations of the cytoplasm may play a role, though certainly a minor one, in the adaptive evolution of higher plants.

THE NATURE OF MUTATION

Before discussing the role of mutation in evolution, we must understand fully the nature of the mutational process, both in respect to the type of change in the genetic material, and the effects of mutations on the phenotype. In its broadest sense, the term, MUTATION, has been used to cover all types of changes in the genes, chromosomes, or other hereditary particles which have a permanent effect on the genotype. Geneticists have, however, generally used the term only for changes at individual loci, which are called gene or point mutations. As our knowledge of gene mutations and chromosomal alterations has increased, different functions of the two types of changes have come to be recognized. Gene mutations have exclusively the effect of adding to the number of alleles available at a locus, and so increasing the gene pool. Chromosomal changes, on the other hand, though they sometimes produce additional genetic variation, have chiefly the effect of changing the linkage relationships between genes, causing certain pairs or larger groups of genes at different loci to be inherited partly or entirely as single units. Consequently, they will be discussed in the next chapter on the organization of genetic variability.

Point mutations can now be defined on the basis of the molecular structure of the hereditary material, or deoxyribonucleic acid (DNA). Since a detailed description of this basic molecular system of heredity is given in every modern textbook of biology or genetics, only those points most essential to our discussion will be brought up here. The molecule of DNA is a double ribbon containing thousands of units, or nucleotides, arranged in a linear order. These are of four different kinds, distinguished by the presence of one of four different bases; the two purines, adenine and guanine, and the two pyrimidines, thymine and cytosine. When forming part of a DNA molecule these nucleotides are usually paired, since the molecular structure of adenine enables it to form a supra-molecular complex with thymine, while guanine can likewise unite with cytosine (Figure 2-3).

The specificity of a DNA molecule, or of any part of it, depends upon the order of these base pairs in the ribbon. Because this specificity is affected by the symmetry of the molecule, the pair adenine-thymine is significantly different from thymine-adenine. Consequently, the DNA molecule contains an informational "code," similar to the Morse code of telegraphy, in which the four nucleotide pairs containing the bases A-T, T-A, G-C, and C-G correspond to two kinds of "dots" and two kinds of "dashes." A "triplet" of three nucleotide pairs determines indirectly, through the medium of messenger ribonucleic acid (*m*-RNA), the position of a particular amino acid in a molecule of a protein.

Proteins have two functions. Most of them, the enzymes, catalyze the numerous chemical reactions which make the organism alive. Others, the structural proteins, determine the form and properties of various parts of living cells, particularly their membranes. For both of these functions, the linear order of amino acids is of primary significance. The specific properties of a particular protein molecule depend upon the order of the twenty kinds of amino acids in the chain-like molecule of which it consists.

Once we understand these facts we can see that any change in the order of nucleotide pairs in the DNA molecule will bring about a corresponding change in the order of amino acids in the protein which it codes. It will greatly alter the properties and functions of the protein, and consequently of the cell of which it forms a part. A point mutation is such a change. Probably the most common type of point mutation is the substitution of one nucleotide pair for another, but deletions or various kinds of transpositions of one or a few nucleotide pairs would also represent point mutations. Most of these latter types of mutations would, however, be deleterious since they would be likely to interfere with some vital metabolic reaction.

Because the battery of enzyme systems which the organism produces during its life is responsible for all of its activities, we would expect to find

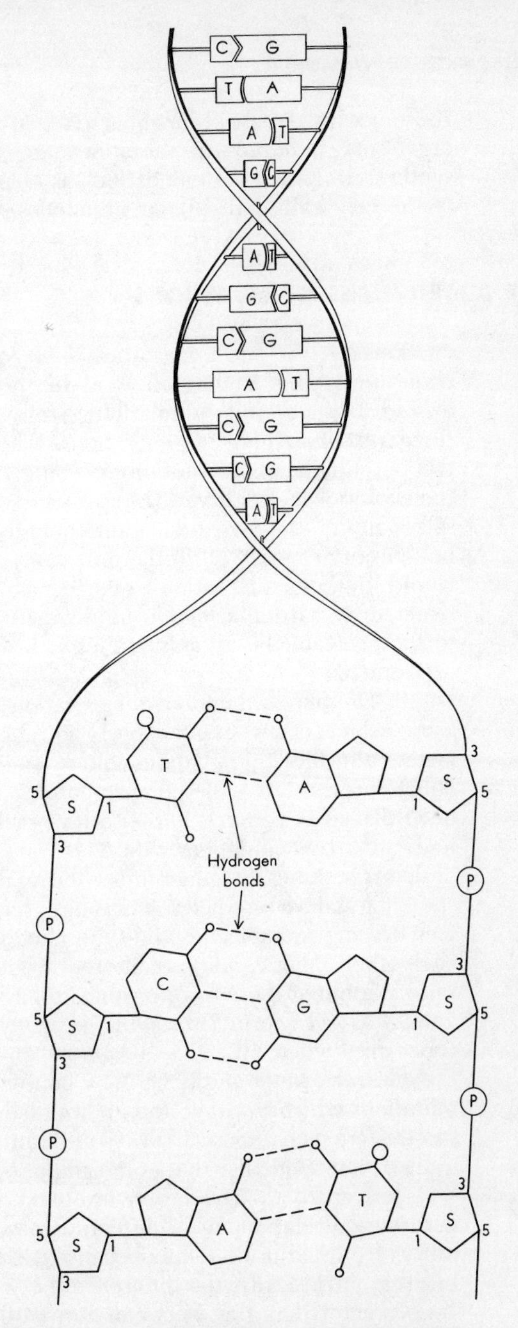

Fig. 2-3. Diagram showing the generalized structure of part of a molecule of DNA, the way in which the order of base pairs in the molecule is duplicated, and the way in which coding of messenger RNA determines the amino acid sequence of proteins. From Bonner and Mills, *Heredity* (Prentice-Hall, Inc.).

among a large number of them, mutations which affect every character in a variety of ways. This is exactly what we find. In the fly, *Drosophila*, one of the best known organisms genetically, mutations are known which affect the color and shape of the eyes, the body color, the size of the fly, size and shape of wings, number and distribution of bristles on the body, number and structure of appendages, and many others. In various higher plants, mutations are known which affect the size, shape and lobing of the leaves, the structure of the inflorescence, the shape and color of the flowers, earliness of flowering, resistance to disease, branching pattern of the plant, and so on. In microorganisms, hundreds of mutations are known which alter or inhibit the synthesis of basic compounds such as amino acids, organic acids, components of the cell wall, and cellular organelles.

THE ADAPTIVENESS OF MUTATIONS

Perhaps the most important fact for students of evolution to realize about mutations is that all modern species and races of organisms have existed as successful populations, well adjusted to their environment, for thousands or millions of generations. We would expect, therefore, that all of the potentially useful mutations would have occurred at least once during the evolutionary history of the species and have been incorporated by natural selection into the gene pool. Consequently, the theoretical expectation would be that all or nearly all of the mutations occurring in a successful population would lower its adaptation to its accustomed environment, and so would be rejected by natural selection unless the environment were changing relative to the needs of the organism. This is, in fact, what has been found in actual experiments.

In *Drosophila*, hundreds of mutations have been obtained by looking for those which occur spontaneously in laboratory cultures as well as by subjecting the flies to radiations and chemical substances which have specific effects on the DNA of the chromosomes. Scores of these mutant flies have been placed in competition with their wild-type alleles in laboratory bottles under standard conditions with nearly always the same result. After a greater or lesser number of generations the mutants are eliminated by the corresponding wild-type alleles. There are, however, a few experiments in which flies bearing mutant and wild-type alleles have been made to compete with each other under conditions different from those under which the fly is usually raised. Some of these have produced different results. For instance, a mutant known as *eversae* in *Drosophila funebris*, is less successful than the "wild-type" flies when raised at 16° or 29°C, but is better than wild-type when the flies are kept at 25°C. In Chapter 4 several examples are given of mutations which improve the ability of the organism to withstand poisonous substances, such as insect sprays and antibiotics.

Further evidence that some mutations may increase the adaptation of their bearers to a particular environment has been obtained from experiments with agricultural plants. One could reason logically that a cultivated field differs from natural habitats because it is man-made and supported by man, but that intrinsically, the differences between it and the surrounding wayside is no greater than that between two natural habitats. Consequently, if some mutations will make a plant more productive in the field, among the great variety of mutations which appear either spontaneously or following radiation or chemical treatment, there should be mutations which will make the plant more productive in a habitat not controlled by man.

For the past twenty years, hundreds of mutations have been produced and tested in cultivated barley. Most of them, as expected, are deleterious. Nevertheless, a small proportion, which has not been carefully estimated but is certainly less than one per cent, has increased productivity. Some mutations have given rise to other valuable characteristics of the barley plant, such as stiff stems which will support the weight of the ripening grain. These mutations, however, do not improve the plant under all conditions in which it might be grown but only under those in which they have been tested and produced. For instance, the best mutations produced by Swedish geneticists are those characterized by early heading and stiff stems. In the moist climate of northern Europe stiff stems are particularly valuable since the moisture which accumulates on the plant during rainstorms tends to make the stems bend over and lie on the ground. In the dry climate of the western United States this happens less commonly, but the ripening grain is often exposed to hot, dry winds. If the stems are too stiff and brittle, these winds will break up the heads and cause the grain to be scattered over the field before it can be harvested. In the language of the agronomist the mutations which are valuable in Sweden because of resistance to lodging are disadvantageous in the United States because of susceptibility to shattering.

This example plus many others tell us that we cannot expect any single mutation to improve the overall fitness of an organism, but only its adaptation to particular environments. Overall improvements in fitness do occasionally occur during the evolutionary history of a group and will be discussed in Chapter 7. The conclusion reached there can be anticipated here. This is that such improved structures have evolved gradually, presumably through the accumulation of many mutations and gene combinations. Furthermore, early stages in the evolution of such structures are found in animals specialized for particular habitats and arise as specific adaptations. Their general value appears only later.

THE RELATIVE IMPORTANCE OF MUTATIONS WITH LARGE AND WITH SMALL EFFECTS

A question which geneticists often ask about the role of mutations in evolution is the following: Granted that some mutations have very great effects on the phenotype while the effects of other mutations are barely perceptible; which mutations have played the largest role in evolution, those with conspicuous effects or those with small effects? This question can be answered only by indirect means. A theoretical answer was given many years ago by the English biometrician, R. A. Fisher, in his well-known book, *The Genetic Theory of Natural Selection*. If an organism is well adjusted to its environment, slight changes in its genetic makeup may adjust it better to modifications of that environment, but drastic alterations of one or a few characteristics are almost certain to make it function more poorly under any environment. The organism might be compared to a racing car carefully tuned up for making the most speed under a given set of conditions. Given a change in temperature, moisture, or condition of the track better performance might be assured by slight adjustments of the carburetor, the transmission, or the fuel, but any radical changes of these elements would certainly reduce the efficiency of the motor and might prevent it from running at all.

Indirect experimental evidence to show that most differences between natural populations have originated by the occurrence and establishment of many mutations, each one with small effects, can be obtained by making hybrids between differently adapted populations and studying their progeny. If segregation in the F_2 generation from such hybrids is according to simple Mendelian ratios, then we can assume that the adaptive characteristics by which they are differentiated are controlled by a small number of genes. Consequently they probably arose through the occurrence and establishment of a small number of mutations, each with a large effect on the phenotype. If, on the other hand, segregation for a particular character is not according to a well defined ratio but forms a blending pattern with most of the F_2 progeny falling at or near a mode intermediate between the parents, then we can infer that many genes contribute to the difference. It has, therefore, originated gradually through the accumulation of many mutations with each individual mutation having a small effect on the phenotype.

Many examples are now available of hybrids between different races in species of both animals and plants. One of the most extensive examples, which is quite characteristic of the usual situation, is that which the botanists of the Carnegie Laboratory at Stanford, California, Drs. Clausen and Hiesey, carried out on the Californian species related to the cinquefoils, *Potentilla glandulosa.* They studied two large F_2 progenies of hybrids between alpine and lowland races of the species which differ in respect to a large number of characters of both vegetative and reproductive organs. Most of the character differences between these races, such as winter dormancy, leaf size and shape, stem height, branching, hairiness, and seed weight (Figure 2-4) are determined by so many gene pairs that their number could not be estimated.

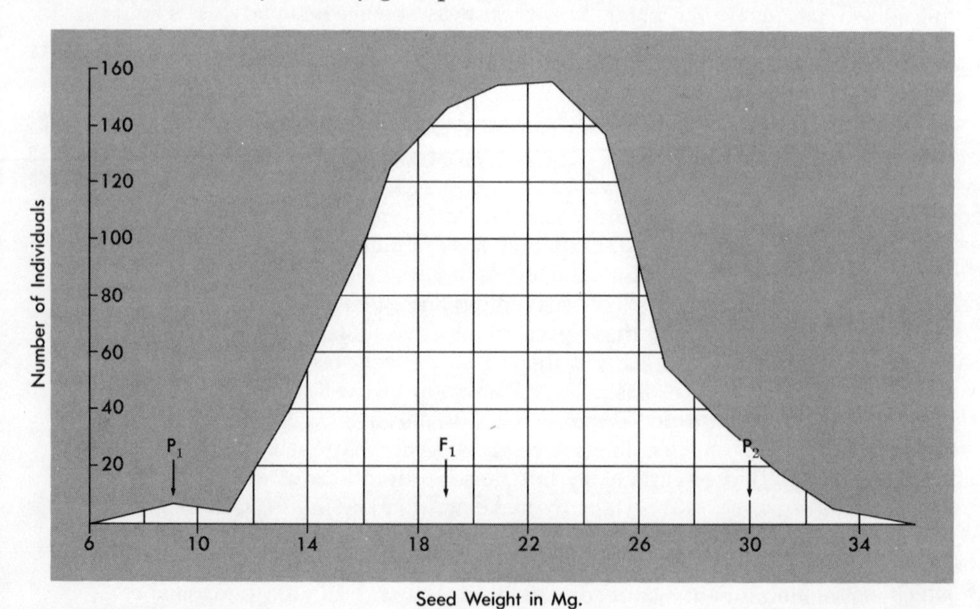

Fig. 2-4. Diagram showing the distribution of mean seed weights in two subspecies of the sticky cinquefoil, one from coastal Southern California and the other from timberline in the high Sierra Nevada, their F_1 hybrid, and a progeny of 961 F_2 plants. Note that the value for the first generation hybrid lies about half way between that for the two parents and that the modal value for the F_2 progeny transgress those for both parents. From Clausen and Hiesey, Carnegie Institute of Washington, Publication No. 615.

Others, such as the color and shape of the petals, are determined by two, three, or four gene pairs. The only possible inference from these data and many others like them is that the races of *Potentilla* diverged gradually through the occurrence and accumulation of a large number of genetic differences. In Chapter 5 evidence will be presented to indicate that the origin of distinct species has also been determined to a large extent by the accumulation of genetic differences with small effects.

There are, however, some conspicuous exceptions to this rule. Some of the best examples involve the origin of mimicry in animals and of different adaptations to pollination in plants. These will be discussed in Chapter 4. They indicate that although the establishment in populations of mutations with conspicuous effects is a very rare event compared to evolutionary change through the accumulation of small differences, nevertheless these rare events may be of great importance to evolution, far out of proportion to the rarity of their occurrence. They may occasionally orient the evolutionary line into an entirely new direction of adaptation.

THE CAUSES OF MUTATIONS

A question often asked is, "What causes mutations?" In laboratory and garden populations kept under normal conditions for the organism in question they occur spontaneously with no apparent cause. Rates of mutation vary widely from one gene locus to another but are always low. Direct experiments to determine the cause of "spontaneous" mutations are almost impossible due to the low rate of their occurrence. Several kinds of radiations, such as x-rays, ultra violet rays, and neutrons have long been known to increase the frequency of mutations at rates proportional to their dosage. This has led some biologists to speculate that some of the "spontaneous" mutations are produced by natural radiations, such as cosmic rays. Yet when calculations are made, based upon known intensities of natural radiations compared to intensities needed to produce mutations, the results show that even where they are most intense, such as at high altitudes in the mountains, the intensity of natural radiations is so low that it could not account for more than a tiny fraction of the "spontaneous" mutations.

Recent studies of the chemical structure of DNA, the results of which have been outlined earlier in this chapter, have led to a new theory of the cause of mutations. This is well described by Dr. F. W. Stahl in his book, *The Mechanics of Inheritance.* The four bases (adenine, thymine, guanine, cytosine) which make up DNA can temporarily exist in rare states which have an affinity for the wrong base and form mismated base pairs. Thus, a rare state of adenine will mate with cytosine instead of thymine (Figure 2-5). When a DNA molecule containing such mismated pairs occurs in a cell, it duplicates to give rise to one strand of normal DNA and one mutant strand, in which an adenine-thymine base pair has been replaced by one of guanine-cytosine (Figure 2-6). Experiments with both coli bacterium and bacteriophage have shown that many genetically identified mutants represent such substitutions of single base pairs.

Strong support for this theory is derived from the fact that a structural analogue of thymine: 5-bromouracil, induces large numbers of mutations in bacteriophage. This base molecule can become incorporated into DNA

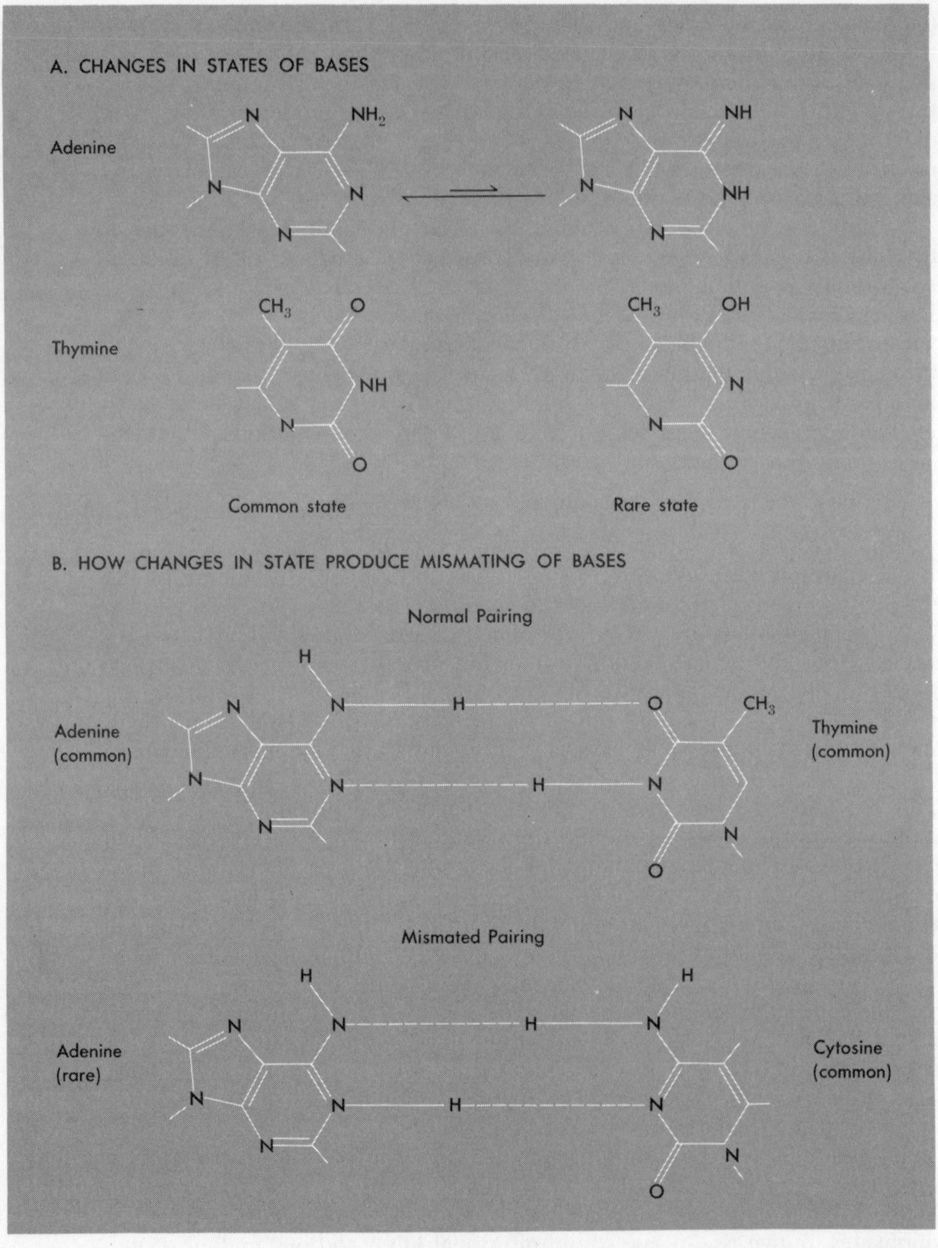

A. CHANGES IN STATES OF BASES

Adenine

Thymine

Common state

Rare state

B. HOW CHANGES IN STATE PRODUCE MISMATING OF BASES

Normal Pairing

Adenine (common)

Thymine (common)

Mismated Pairing

Adenine (rare)

Cytosine (common)

Fig. 2-5. *Upper left* and *right,* normal and tautomeric formulae for adenine; *second row,* the same for thymine; *third row,* normal pairing of adenine and thymine in a typical DNA molecule; *fourth row,* abnormal pairing of tautomeric state adenine with cytosine. Modified from Stahl, *The Mechanics of Inheritance* (Prentice-Hall, Inc.).

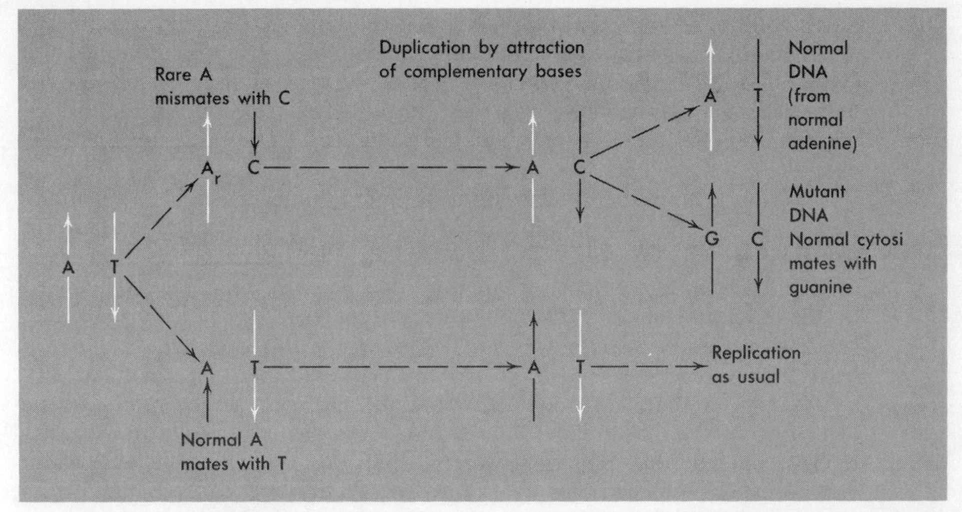

Fig. 2-6. Diagrammatic sequence showing how change of adenine from the normal to the tautomeric state can produce substitution of a G-C base pair from an *A-T* pair, this constituting a mutation. Modified from Stahl, *The Mechanics of Inheritance* (Prentice-Hall, Inc.).

instead of thymine. Chemical evidence suggests that it changes more easily than thymine into a state which attracts guanine instead of adenine and so induces mutation by base pair substitution. If 5-bromouracil induces mutations in this fashion, it should also speed up reverse mutation when cells containing mutations induced by it are again exposed to the chemical. This has proven to be the case.

Mutation by such rare states of base pairs may be only one of several ways in which slight chemical instabilities of cellular organization could produce alterations of molecular structure in chromosomal DNA and the compounds associated with it. Perhaps the safest generalization to make at present is to state that "spontaneous" mutations are by-products of accidents in nuclear or even cellular metabolism.

RATES OF MUTATION AND OF EVOLUTION

As was already mentioned in the first chapter, and will be emphasized in the next, mutations are rarely if ever the direct source of variation upon which evolutionary change is based. Instead, they replenish the supply of variability in the gene pool which is constantly being reduced by selective elimination of unfavorable variants. Because in any one generation the amount of variation contributed to a population by mutation is tiny compared to that brought about by recombination of pre-existing genetic differences, even a doubling or trebling of the mutation rate will have very little effect upon the amount of genetic variability available to the action of natural selection. Consequently, we should not expect to find any relationship between rate of mutation and rate of evolution. There is no evidence that such a relationship exists. Studies of mutation rates are of great importance in increasing our understanding of genetics, but they are of relatively little importance to the evolutionist. To understand the basis of differing evolutionary rates, we must learn as much as possible about the interrelationships between populations of organisms and their environment. This topic will be discussed in later chapters.

We must, however, ask the following question: has a scarcity of suitable mutations ever been a limiting factor in evolution? A few simple statistics will help us to answer this question. While rates of mutation vary enormously from one gene locus to another, and can also be greatly influenced by the environment, a rate of one mutation per gene locus in every 100,000 sex cells is a conservative estimate. Because all higher organisms contain at least 10,000 gene loci, and most of them contain many more, we can conservatively say that one individual out of ten carries a newly mutated gene at one of its loci. As already pointed out, the great majority of these mutations are deleterious, but a small proportion of them are beneficial. From various experimental studies we can arrive at a conservative estimate of the proportion of useful mutations as one in a thousand. On the basis of these estimates we can calculate that in any species about one in ten thousand individuals in each generation would carry a new mutation of potential value in evolution. Using conservative values of 100 million as the total number of individuals per generation and 50,000 as the number of generations in the evolutionary life of the species, we could expect that at least 500 million USEFUL mutations would occur during this life span. We do not know how many new mutations are needed to transform one species into another, but five hundred is a reasonable estimate. On this basis, only one in a million of the useful mutations or one in a billion of all mutations which occur needs to be established in a species population in order to provide the genetic basis of observed rates of evolution.

This final calculation is based upon the assumption that any of the rare useful mutations which occurs is capable of being combined with other useful genes and spread through the species population by means of natural selection. Given the opportunity of free intercrossing, gene exchange, free recombination of genes among all of the individuals of a species, and the number of generations which occurs during the long evolutionary life span of a species, this is theoretically possible as the calculations of J. B. S. Haldane and Sewall Wright showed us many years ago. Actually, however, free gene exchange and recombination between all members of a species is never possible. In a widespread species the long distances which separate individuals in different parts of its range are themselves a partial barrier. Often instinctive preferences for mating between individuals which live near each other provide additional barriers to gene exchange.

MUTATION, RECOMBINATION, AND ADAPTATION

Free recombination between genes is also restricted by a number of factors inherent in the structure of natural populations and the way in which genes are organized in the chromosomes of the cell nuclei. These factors will be discussed in the next chapter. Our calculations lead us, therefore, to the conclusion which has been reached by most geneticists who are studying evolutionary processes. The chief limiting factor on the supply of variability for the action of natural selection is not the availability or rate of occurrence of mutations, but the restrictions on gene exchange and recombination which are imposed by the mating structure of populations and the structural patterns of chromosomes. Natural selection directs evolution not by accepting or rejecting mutations as they occur, but by sorting new adaptive combinations out of a gene pool of variability which has been built up through the combined action of mutation, gene recombination, and selection over many generations. For the most part Darwin's concept of DESCENT WITH MODIFICATION fits in with our modern concept of interaction between evolutionary processes, because each new adaptive combination is a modification of an adaptation to a previous environment. Although mutations may occasionally play an important role in directing natural selection along a particular channel, most of them are important only as contributors to the gene pool. Consequently, the rate of mutation rarely if ever has an influence on the rate of evolution.

Mutations could have a direct influence on the rate and direction of evolution only if they occurred on an essentially uniform genetic background, i.e., in a population of individuals genetically alike and homozygous for the most important genes concerned with adaptation. As suggested earlier in this chapter and to be brought out clearly in the next chapter, populations of sexually reproducing and cross fertilizing organisms are never genetically homogeneous. If they were, they would lose their ability to evolve in response to changes in the environment. The principal reason for this is the extreme complexity of adaptive systems in higher organisms. This is due principally to three factors. The first of these is the highly complex environment. It includes both the physical surroundings of a population such as climate, topography, nature of available water and soil, and the numerous other kinds of organisms with which a population is inevitably associated. Furthermore, most environments are constantly changing in a more-or-less regular fashion. In temperate climates the seasonal differences between summer and winter are the most obvious of these changes, but seasonal differences in rainfall are found in many parts of the tropics as well. Cyclic changes in the abundance of different kinds of organisms are well known, as are changes in the abundance of the flowers and fruits of trees upon which many small animals depend for their subsistence. Longer-lived organisms must have enough flexibility in their individual bodies to cope with these variations, while shorter-lived species such as the smaller insects must have populations which can shift their genetic makeup back and forth as the environment changes.

Secondly, the integration of the body itself is a highly complex affair involving precise adjustments between processes and functions which are controlled by completely different genes. The adaptation of a reptile or mammal to a diet of smaller animals involves specialization of the limbs for seizing and holding its prey; of the jaws and teeth for tearing, chewing, and swallowing it; and of the digestive system for digesting animal proteins. If such an animal is to produce by evolution a new line of organisms adapted to feeding exclusively on leaves, all of its parts must be changed synchronously and in harmony with each other to fit each of them to its new and completely different role. The number of necessary replacements and adjustments of parts would be as great as those which would be needed to convert a propeller driven airplane into one with jet propulsion.

The third basic cause of the complexity of gene-controlled adaptations is the complexity and indirect nature of the relation between the gene and the character. As can be learned from any modern textbook of genetics, the immediate function of genes is to provide an informational code for the molecular structure of proteins. Some of these, the structural proteins, determine the supramolecular structure of the cytoplasmic organelles and the cell wall. Others, probably the great majority of gene-produced proteins, are enzymes whose molecular configuration is responsible for the rates and conditions of operation of the numerous metabolic processes going on in the cell. Some of the adaptive adjustments associated with evolution can be produced directly by gene mutations which produce altered properties. Adaptations of plants to new climatic regimes provide good examples. On the other hand, the structural modifications which must accompany evolutionary changes in an animal's diet must involve alterations of the developmental patterns of whole tissues and organs, and so require coordinated changes in the activity of many different enzymes. Furthermore, if the action of a single enzyme is altered, many different characteristics of the adult organism can be changed in different ways. For instance, mutation of a gene responsible for one of the enzyme-controlled processes needed to produce cartilage in a higher animal can change bone structure, muscular movements, breathing, and many other activities. Some of these changes could be beneficial, others deleterious. If such a mutation were to take part in the evolution of a new adaptive system, it would have to be combined with other mutations which would counteract or suppress its harmful effects while leaving unchanged or accentuating its beneficial effects.

Two examples from the recent literature on genetics will serve to illustrate the complexity of the gene interactions needed to produce even the simplest of structures. The chloroplast of higher plants is of microscopic dimensions (5–8 μ in diameter) and has a similar makeup in all kinds of plants, from ferns to conifers, oak trees, daisies, and grasses. It consists of an outer membrane which surrounds a large number of inner membranes or lamellae. The lamellae stretch across the chloroplast in parallel fashion and have a regular alternation of thicker and thinner regions (Figure 2-7). The lamellae themselves are a regular, intricate fabric made up of molecules of chlorophyll, proteins, and lipids.

As mentioned earlier in this chapter, many artificial mutations have been produced in cultivated barley by X rays, ultraviolet, and chemical mutagens. By far the most common mutations are those which affect the structure and development of chloroplasts. Based upon his extensive knowledge of these mutations, Dr. Åke Gustafsson of Sweden has estimated that from 250 to 300 gene loci in barley are concerned with the synthesis of chloroplasts. One of Dr. Gustafsson's associates, Dr. Diter von Wettstein, has studied chloroplast development in several of these mutants and has found that different mutants affect very different developmental stages and processes. Chloroplast development consists of a long sequence of gene controlled processes which must be coordinated so as to follow each other in a

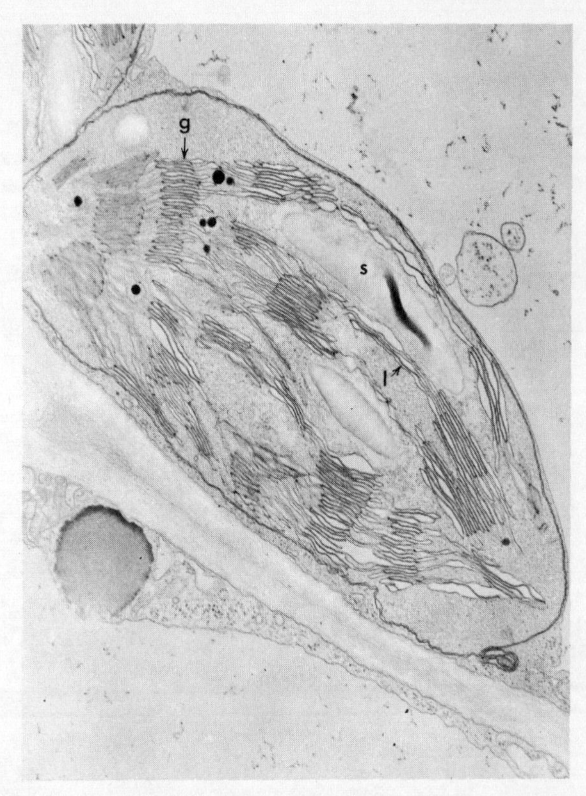

Fig. 2-7. Fine structure of the chloroplast of a higher plant, as seen with the aid of the electron microscope. From Galston, *The Life of the Green Plant* (Prentice-Hall, Inc.).

precisely integrated fashion (Figure 2-8).

Genetic studies of another organelle, the flagellum of the unicellular microorganism *Chlamydomonas*, have shown that its development depends upon a similar complex sequence of gene controlled processes. Dr. Ralph Lewin has found mutations affecting flagellar structure at twelve different loci. These alter the adult structure and impede the normal activity of the flagellum in a variety of different ways.

In order to produce adaptive changes of such complex, highly integrated developmental systems, many coordinated alternatives of the genotype are necessary. The next chapter will explain how the gene pool is organized so as to make genetic changes in such adaptive systems possible.

Chapter Summary

Phenotypic variability in populations has three components: (1) environmental modification, (2) genetic recombination, (3) mutation. The great bulk of the variation found in any natural population is contributed by (1) and (2). Environmental modification is greatest in organisms subjected to violent

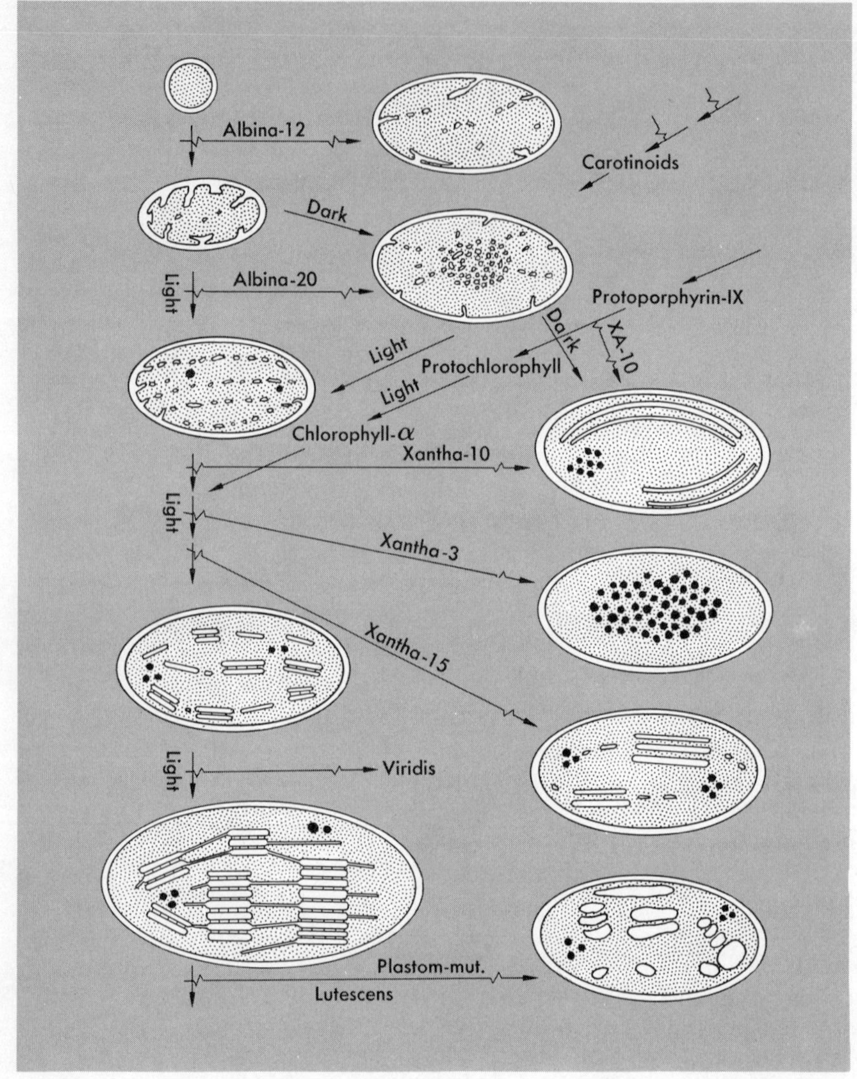

Fig. 2-8. Stages in the development of the chloroplast of barley showing how it can be influenced by deleterious mutant genes. From von Wettstein, *Canadian Journal of Botany,* 39:1537, 1961.

fluctuations in their environment, and is greatest in respect to characteristics which are accessory to rather than characteristic of the basic processes of metabolism. In higher animals, the separation of the germ cells from the body cells and tissues at an early stage in embryonic development provides a strong theoretical argument against the hypothesis first advanced by Lamarck, that adaptive phenotypic modifications acquired during the lifetime of an individual can be converted into hereditary differences and so contribute to evolution. In plants and microorganisms this separation does not exist, but experimental evidence in them also runs counter to the hypothesis that the inheritance of acquired modifications is a major factor in evolution. A further theoretical argument against this hypothesis is the complexity and indirectness of the network of interactions which are interposed between the primary action of the genic DNA and the final expression of the phenotypic character.

Although the great bulk of new mutations are inadaptive, a small proportion of adaptive or potentially adaptive mutations has been found in those organisms which have been most intensively studied genetically. Evidence from hybridization between genetically different natural populations has led to the conclusion that most of the mutations which have contributed to evolution have been those which individually have small effects on the phenotype. Spontaneous mutations are probably caused by irregularities in replication of the DNA molecule, as well as other accidents in nuclear or cellular metabolism.

Rates of mutation and of evolution are weakly if at all correlated with each other. This is because in any one generation, mutation contributes very little to the gene pool of variation and is at random with respect to the direction of adaptation. Calculations based upon known mutation rates, sizes of populations, and the evolutionary life of races and species in terms of generations show that mutation can provide an ample source of variability to replenish the gene pool as it is depleted by selection even if only a small fraction of the potentially useful gene mutations are incorporated into adaptive gene combinations. Because adaptation in higher organisms is based upon complex, highly organized combinations of genes, genetic recombination is a necessary intermediary between mutation and selection.

Questions for Thought and Discussion

1. In what ways can environmental modification of the phenotype contribute to the adaptive evolution of populations, and in what ways is it unable to do so? Give as many reasons as possible for your answer.

2. Why are most mutations inadaptive? Can you imagine a situation in which a population might find itself, which would increase the number of potentially adaptive mutations?

3. Is there any reason for believing that an extra supply of energy is needed to produce mutations at a rate sufficiently rapid to permit evolution to take place? What bearing does your answer have on a question which physical scientists have sometimes raised: the relationship between the evolutionary tendency toward greater complexity in many lines and the second law of thermodynamics, or entropy?

4. Using examples, explain as fully as you can why there is little or no correlation between rates of mutation and rates of evolution.

CHAPTER 3

The organization of genetic variability in population

In the last two chapters, the statement has been made that all cross breeding populations possess a gene pool of variability consisting of two or more different alleles at many of their gene loci. If this is so, then we would expect that all of the individuals of these populations would be heterozygous at a large number of loci, i.e., they would possess different alleles at the same locus on a pair of homologous chromosomes, one derived from their father and one from their mother. Furthermore, either artificial or natural selection could change such populations by sorting out those individuals having a high proportion of positive alleles for the character being selected. Through crossing between these selected individuals followed by further selection, all of the positive alleles, formerly scattered through the population, would become concentrated in a few individuals. These individuals would then possess the character to a high degree. All this could happen without the occurrence and establishment of a single new mutation.

These theoretical expectations have been realized in a number of experiments. The appearance of unusual phenotypes is a common result of experiments in which close inbreeding, such as mating between brothers and sisters, is carried out for several generations on a normally outcrossed population. The genetic explanation for the appearance of these unusual phenotypes is Mendel's first law of segregation. If the phenotype is produced by a recessive gene, *a*, it will appear in one fourth of the progeny derived from a mating between two heterozygotes, $Aa \times Aa$. In these parents, the trait is inhibited by the action of the dominant allele *A*, so that neither parent possesses it.

Several experiments with cross fertilizing populations have shown that they can respond to selection for quantitative characters for many generations, while still retaining a considerable amount of genetic variability with respect to the character in question. One of the best known of these experiments was begun in 1895 by agronomists at the University of Illinois who decided to find out how long they could obtain a response to selection for certain characteristics of a normal open pollinated crossbreeding population of field corn. They selected for four characteristics of the kernels: high and low protein content, high and low oil content. This experiment was continued for fifty generations. In the case of high oil and low protein, the population responded significantly even between the forty-fifth and fiftieth generations. During the experiment, the protein was more than doubled in the high-protein line and reduced to less than half of the original concentration in the low-protein line. Even greater changes were obtained by selecting for high and low oil content. The results of the experiment are shown in Figure 3-1. After the experiment was concluded, reverse selection was practiced successfully on both the populations having high protein content and those with a high oil content. This shows that the selection did not exhaust the gene pool with respect to genes for the character in question.

The facts which we know about mutation rates in corn tell us that the Illinois agronomists must have been sorting out (at least in part) differences which existed in their original population. Since in each line of corn being selected the number of plants raised per generation was between 200 and 300, the total number of plants raised during fifty generations in each selection line was between 10,000 and 15,000. We do not know the actual rates of mutation in corn for changes in oil or protein content of the kernels. But data on rates of mutation in respect to other quantitative characteristics in plants tell us that for a particular characteristic the occurrence of one mutation in 5000 plants is a relatively rapid rate. Hence the occurrence of many mutations in the desired direction during the course of selection is rather unlikely.

Experiments such as these have led geneticists to recognize that the amount of variation which can be seen in a population is only a fraction of the genetic variability which exists in it. All crossbreeding populations contain a hidden store of variability in the form of recessive genes and gene combinations which can be brought to the surface by inbreeding and selection. Two explanations have been given for this fact. One is that there is always a lag between the occurrence of an unfavorable mutation and its elimination by natural selection. If the mutation is recessive, individuals heterozygous for it will be perfectly normal in appearance and will not be eliminated by selection. If such a mutation happens to occur at a position on a chromosome close to some gene which has a particularly high selective value, the undesirable recessive will be carried through the population through association by genetic linkage with its valuable neighbor until the two are separated by genetic crossing over. There is plenty of evidence to show that this "mutational load" of undesirable recessive alleles exists in natural populations. When wild populations of *Drosophila* are inbred, a large number of lethal or semilethal alleles are uncovered. In respect to many of them, heterozygous individuals have been obtained and compared with individuals containing only wild-type alleles at the locus in question. Such comparisons have shown in numerous instances that the heterozygous flies are indistinguishable in appearance from homozygous normal flies, and compared to them have the same or a lower selective value.

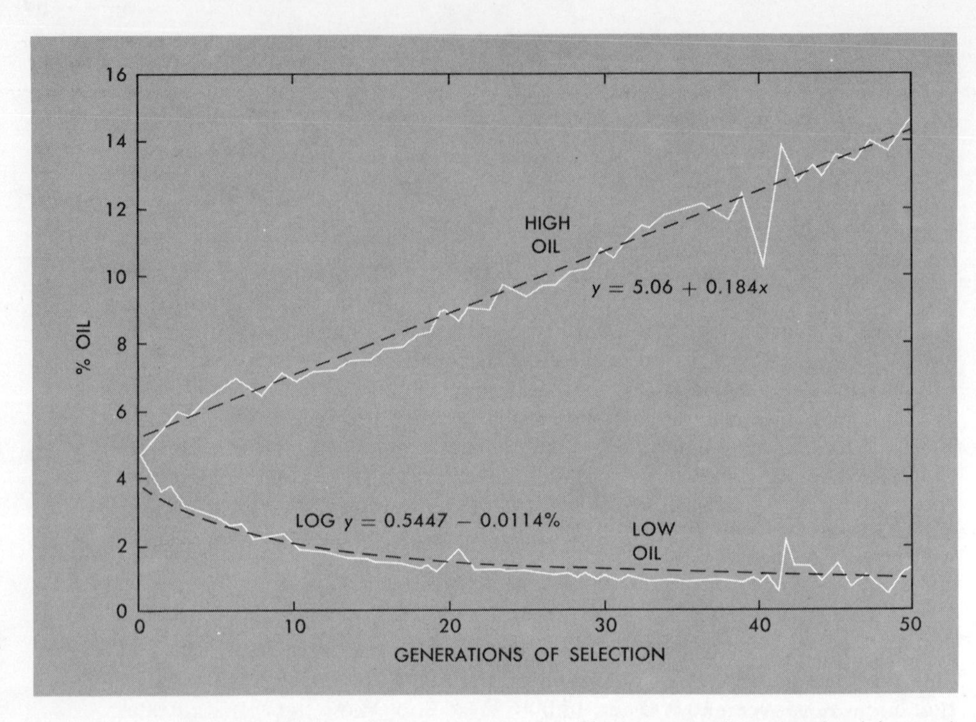

Fig. 3-1. Response to artificial selection over fifty generations for increased and decreased protein content and for similar changes in oil content of the kernel, starting with an open pollinated variety of field corn. From Woodworth, Jugenheimer and Lenz, *Agronomy Journal* 44:60, 1952.

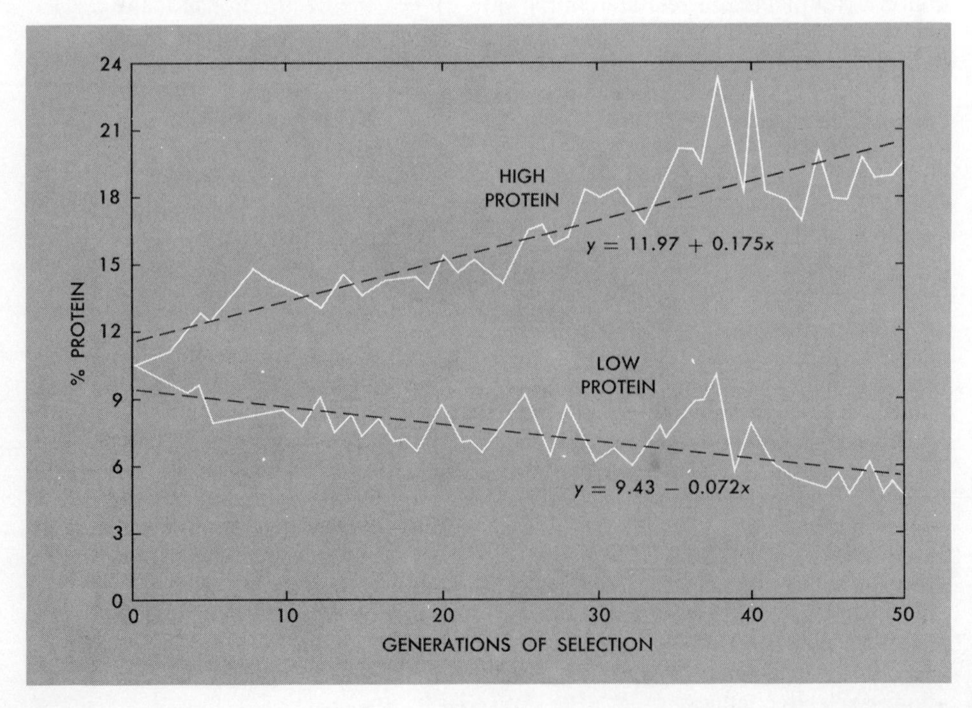

Such results, however, have by no means been the only outcome of experiments of this sort. In many instances, individuals which were heterozygous for certain chromosomal combinations have proved superior to those homozygous for either of the parental chromosomes. To understand the experiments which have led to such conclusions, we must remind ourselves of some of the facts of population genetics. If two alleles, *A* and *a*, occur in a population in equal frequency, then the result of random mating between the three possible genotypes, *AA*, *Aa*, and *aa*, will be to maintain them continuously at the frequencies which are found in the F_2 progeny of the simple monohybrid cross ($AA \times aa$), namely $\frac{1}{4}AA$, $\frac{1}{2}Aa$, $\frac{1}{4}aa$. If, on the other hand, the frequencies of the alleles are unequal, the frequencies of genotypes must be calculated according to the following algebraic formula, which is the simple expansion of a binomial. If the frequencies of the alleles *A* and *a* are represented by the symbols p and q, then the genotypes *AA*, *Aa*, and *aa* will occur in the frequencies p^2, $2pq$, and q^2. This formula, known as the Hardy-Weinberg equilibrium, illustrates a principle which could be called genetic inertia. It tells us that Mendelian segregation by itself cannot lead to changes in gene frequency in populations. Furthermore, the Hardy-Weinberg equilibrium exists only if mating is at random and all three of the genotypes have equal adaptive values. Evolutionary changes in populations must depend either upon unequal or differential adaptive values of genotypes, on deviations from random mating, or on both of these factors acting together. If the three genotypes, *AA*, *Aa*, and *aa*, can be recognized either by their appearance or by progeny testing, then comparisons between their actual frequency and that which is expected according to the Hardy-Weinberg formula can tell us something about the forces which may be acting to change the frequency of genes in a population. Such comparisons are usually made with the aid of the statistical "goodness of fit" test using the value chi square. If this test shows that the relative frequency of the two homozygotes is according to expectation but the heterozygotes are deficient, it indicates either that heterozygotes are less fit than their homozygous parents or that the population is undergoing a significant amount of inbreeding. If one of the two homozygotes is less frequent than expected, this indicates selective pressure against the allele in question. If, on the other hand, the two homozygotes occur at the expected frequencies but the heterozygotes are more frequent than expected, this indicates that the heterozygotes are adaptively superior to either of the homozygotes (Table 3-1). This superiority of heterozygotes is one of the genetic phenomena responsible for heterosis or hybrid vigor. Plant breeders have taken advantage of this phenomenon to produce high-yielding hybrids of corn, sorghum, and other crop plants.

One of the clearest examples of a deleterious gene which is retained in a population because of heterozygote superiority has been found in populations of a species of fly (*Drosophila polymorpha*) in Brazil. As shown in Figure 3-2, populations of this species contain flies with three types of coloration on their abdomen. In any population, the proportions of the three types of flies remain constant for many generations. The genotype (*ee*) which determines the lightest color is always the least frequent, indicating an inferior adaptive value. In artificial populations produced by crossing homozygous dark (*EE*) and light flies, the relative frequencies were those in the following

TABLE 3-1. Relation between Adaptive Values of Genotypes and Persistence of Less Adaptive Phenotypes.

In this table, the symbol $>$ means IS BETTER THAN
the symbol $=$ means IS EQUAL TO
the symbol $<$ means IS POORER THAN
A and a can refer to alleles at individual gene loci, but in actual populations are more likely to represent groups of genes linked together on small, corresponding chromosomal segments.

SITUATION	RESULT
1. $AA > Aa > aa$	Population becomes homozygous for *AA*, and *aa* phenotypes disappear.
2. $AA = Aa > aa$	Constant elimination of *aa* genotypes reduces the frequency of *a*, so that fewer *Aa* genotypes are formed. These eventually become so rare that *Aa* × *Aa* matings never occur, and *aa* phenotypes do not appear unless the population is inbred.
3. $AA < Aa > aa$	Frequency of *Aa* is maintained at a higher level than expected on the basis of the Hardy-Weinberg equilibrium. Repeated matings of *Aa* × *Aa* individuals maintain both *AA* and *aa* phenotypes in the population, even if *aa* is greatly inferior to *AA*.

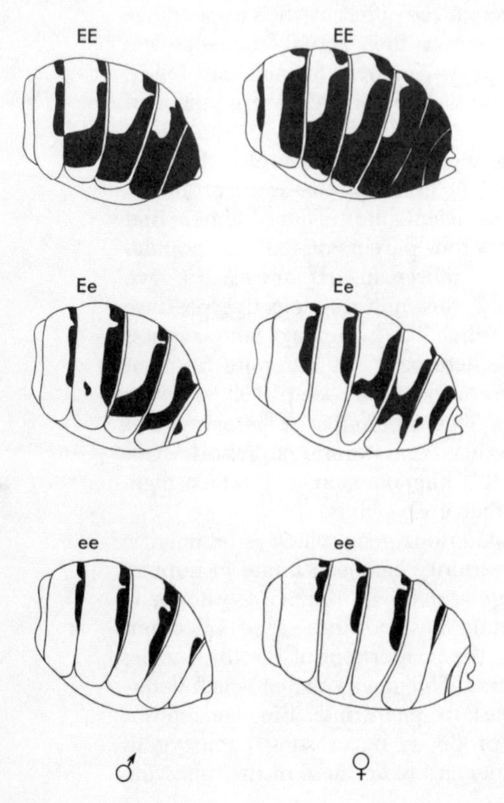

Fig. 3-2. Coloration of the abdomen of two alternative geneotypes of the South American species of fly *Drosophila polymorpha* and of their superior heterozygote. From da Cunha, *Evolution,* 3:239, 1949.

table, where they are compared with values expected on the basis of the Hardy-Weinberg equilibrium (Table 3-2).

We can conclude from these data that flies heterozygous for part of a chromosome bearing the *E* locus have an adaptive advantage over homozygotes for this chromosomal region. We cannot conclude without further tests, however, that the color gene itself is responsible for this advantage. The units of segregation in meiosis are chromosomes, not genes. Recombination between different genes located on the same chromosome depends upon a special process known as crossing over which is described in all textbooks of genetics. If two genes are very close together on a chromosome, crossing over between them is a rare event and may occur in less than one per cent of gametes. As one can see from the chromosome maps of *Drosophila* published in many textbooks of genetics, groups of as many as six to eight gene loci may be so tightly linked together that crossing over between those farthest apart is less than five per cent. Unless they were continued over many generations such low frequencies of crossing over would not have an appreciable effect on genotype frequencies such as those given above. The methods now available to geneticists, therefore, can only in exceptional circumstances distinguish between the adaptive advantage of a single gene locus in the heterozygous condition and that of clusters or blocks of genes which are usually inherited as units because they are located beside each other on the same chromosome.

One of these exceptional circumstances is provided by new alleles which have been produced in the laboratory as mutations induced by radiations. In well-known organisms like *Drosophila* these can be distinguished from changes in chromosomal segments or blocks of genes by appropriate genetical tests. Experiments designed to produce such new alleles which give superior heterozygotes when combined with wild-type alleles have met with some success. Nevertheless, the advantage of these heterozygotes, even when present, has been slight. By far the majority of radiation-induced mutations produce heterozygotes which are no better than and often inferior to homozygous wild-type flies. We can conclude from these experiments that when heterozygotes for a particular chromosome or chromosomal region are superior

TABLE 3-2. Frequency of three phenotypes in a laboratory population of F_2 generation raised in the laboratory from a cross between dark (*EE*) and light (*ee*) individuals of *Drosophila polymorpha* (Data from Da Cunha, 1949).

	DARK (*EE*)	INTERMEDIATE (*Ee*)	LIGHT (*ee*)
Observed	1605	3767	1310
Expected	1670.5	3341	1670.5
Deviation	−65.5	+426	−359.5
Relative adaptive values	0.85	1.00	0.695

to their related homozygotes, this superiority is more likely due to the interaction of alleles at several different gene loci, the phenomenon designated by geneticists as EPISTATIC INTERACTION, than to simple interaction between the dominant and the recessive allele at a particular locus.

The superiority of heterozygotes for either genes or chromosomal segments is important because it maintains an extra supply of genetic diversity in populations. If a population exists in a relatively homogeneous habitat so that all of its individuals are competing directly with each other, two opposite alleles which control very different phenotypes are not likely to be retained in the population for an indefinite length of time. One phenotype will usually have a selective advantage over the other, and the allele controlling it will therefore increase continuously in frequency. If the heterozygote has a lower adaptive value than the homozygote for the superior allele, the inferior allele will be driven out completely. If the heterozygote is equal in adaptive value to the superior homozygote, a situation which usually exists in respect to dominant and recessive alleles at a locus, the recessive allele will remain in the population at a low frequency and will constitute a part of its "mutational load." If, however, the heterozygote is superior to either homozygote, the inferior allele will be retained at a fairly high frequency in the population, even if genotypes homozygous for it are lethal.

Heterozygote superiority of this type is the only explanation for a number of results which have been obtained from studies of natural populations of both animals and plants. As an example, seedlings raised from seed gathered from single plants of diploid orchard grass (*Dactylis glomerata judaica*) in Israel included a large number of albino seedlings which died as soon as the reserve food supply in the seeds was exhausted (Figure 3-3). Albinos are well known in grasses. They are invariably produced by genotypes homozygous for recessive alleles which inhibit the formation of chloroplasts and chlorophyll. Such recessive alleles must have existed in the heterozygous condition in the maternal plants and were transmitted to half of its egg cells. The same alleles must have been in some of the pollen grains which gave rise to the seed, having come from other heterozygous plants. Since orchard grass is self-incompatible and will not set seed with its own pollen, the seed which gave rise to the albino seedlings must have been produced by crossing two parents both of which were heterozygous for the same recessive allele. From this information, the experimenters concluded that the populations of orchard grass to which these plants belonged must have contained from 5 to 48 per cent of individuals heterozygous for the same pair of recessive lethal genes. Since such individuals would usually produce some lethal offspring, and so would have a lower reproductive capacity than individuals free of such lethals, they could be kept in the population only if they were adaptively superior to individuals homozygous for normal alleles. The fact that the percentage of heterozygotes was strongly correlated with the soil on which the plants were growing gives further support to the hypothesis that these plants remained in the population because of their adaptive properties.

One can easily see that if heterozygosity for single gene loci or for groups of genes linked together on the same chromosomal segment gives genotypes an adaptive superiority, many alleles will be retained in the

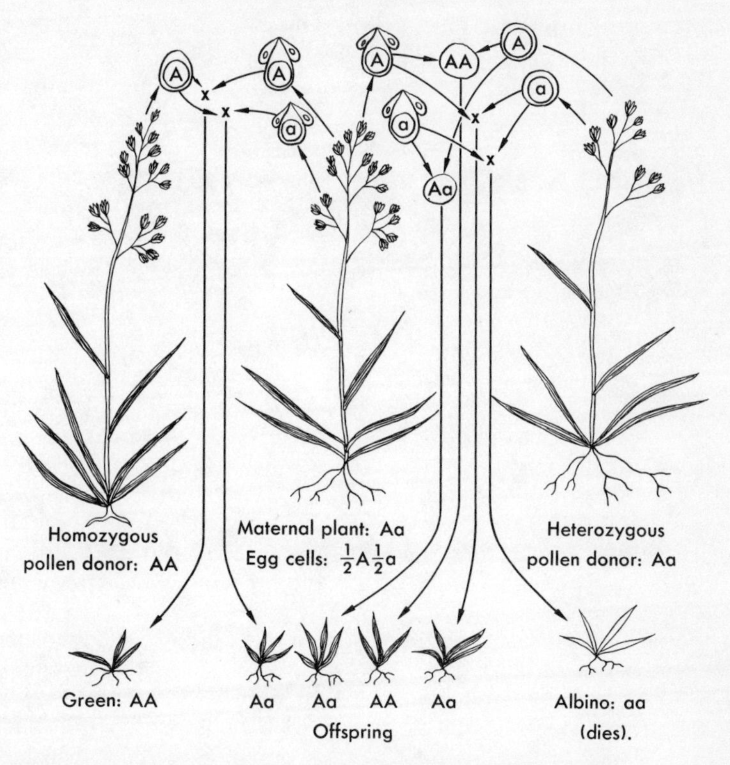

Fig. 3-3. Diagram to show how cross pollination between individual plants of a population which have normal green color but are heterozygous for the same gene responsible for albino seedlings (*a*) can yield occasional lethal homozygotes (*aa*) among the seedlings raised from seed gathered from wild plants cross pollinated in nature. Zygotes are designated either by their genotypes (AA, Aa) or by the symbol x. From data of Apirion and Zohary.

population which would be otherwise eliminated, and the gene pool will be correspondingly enriched. Since the inferior alleles are retained in the population because their effects in the heterozygote are more than counterbalanced by those of the superior alleles, the enrichment of the gene pool by heterozygote superiority is called the BALANCED LOAD of genetic variability. If this load consists of two or more very different and discontinuous phenotypes, it produces a condition known as BALANCED POLYMORPHISM.

The existence of a balanced load is supported by a large amount of experimental evidence, particularly from populations of *Drosophila.* The heterozygosity which produces this balance usually involves large segments of chromosomes rather than individual gene loci. To understand this phenomenon, therefore, we must review the types of changes which chromosomes can undergo above the level of the gene locus and the effects which these changes can produce.

The four possible types of changes are DELETIONS, DUPLICATIONS, INVERSIONS, and TRANSLOCATIONS (Figure 3-4).

The DELETION or removal of a gene locus or a group of loci from a

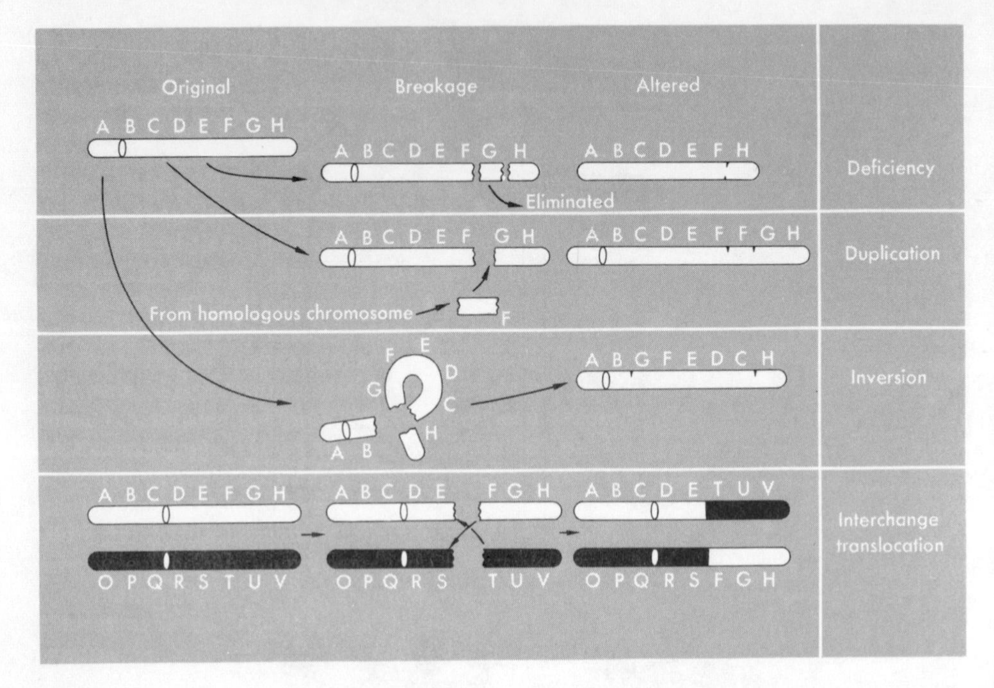

Fig. 3-4. Diagrams to show how chromosome breakage and reunion can give rise to the four principal structural changes which chromosomes undergo: deficiency, duplication, inversion, and translocation.

chromosome is most often lethal in the homozygous condition. Occasionally the function performed by the normal alleles at the missing loci can be taken over by genes in some other part of the chromosomal complement. In other instances the organism may have changed its environment so that a particular chemical synthesis is no longer needed. The first exceptional situation often exists if whole chromosomes have become duplicated in the cells. Losses of chemical syntheses might be expected to occur as organisms pass from an independent existence to a saprophytic or parasitic mode of life, but experimental evidence for this type of evolutionary change is not now available. Consequently, we have no good evidence at present to indicate that deletions of chromosome segments by themselves play a significant role in evolution.

The DUPLICATION of a gene locus may cause an imbalance of gene activity which will reduce the viability of an organism. A well-known example is the BAR duplication in *Drosophila,* which causes the eyes to become abnormally narrow (Figure 3-5). When originally discovered, bar was thought to be a dominant mutation. Later on, when geneticists found that the giant chromosomes in the salivary glands of this fly reflect the pattern of its gene loci, they examined the region where bar was known to be located and found that flies having the phenotype of bar always had a certain number of bands in this region repeated. Flies with a greatly exaggerated bar phenotype had this region present three times.

Other duplications of chromosomal segments in various organisms have

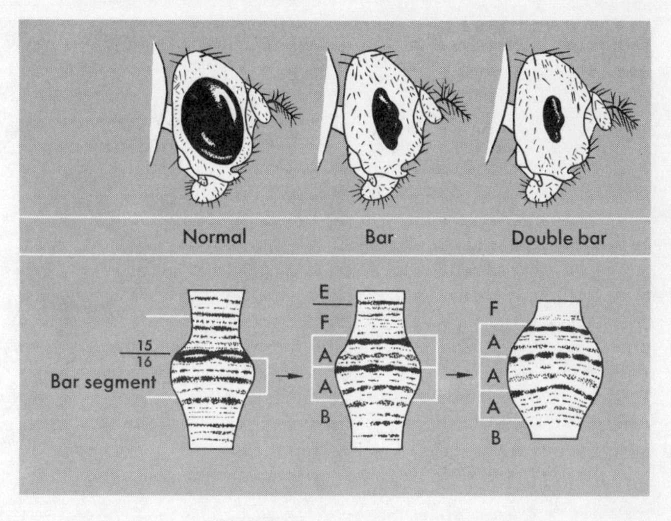

Fig. 3-5. A portion of the *X* chromosome of *Drosophila melanogaster* showing the duplicated pattern of banding which gives rise to the dominant phenotype bar, which is also illustrated. From Gardner, *Principles of Genetics* (John Wiley & Sons, Inc.).

much milder effects on the phenotype. In an annual species of flowering plant, *Clarkia unguiculata*, whole chromosomes may be present in the somatic tissues in triplicate rather than in duplicate without producing any appreciable change in either the appearance or the vigor of the plant. Such TRISOMICS are, however, genetically unstable, and in natural populations will tend to revert by segregation to genotypes with the normal chromosome number.

Since some organisms can tolerate duplications of chromosomal material, duplications may play a role in the evolution of organisms having an increased variety of gene-controlled activities. If a particular gene locus were present in duplicate, then one of the twin loci could mutate to an allele having an entirely different function without disturbing the adaptiveness of the organism, since unchanged alleles at the other locus could adequately perform the original function of the locus.

For the role of the third type of chromosomal change, INVERSIONS, a great wealth of experimental evidence is now at hand. An inversion occurs when a chromosome breaks in two places. The segment between the breaks then becomes turned around, so that the order of its genes is reversed with respect to that on the unbroken chromosome (Figure 3-4). Spontaneous inversions probably occur through accidental breakage at the prophase of either mitosis or meiosis. At these stages the chromosomes are long and slender, are actively moving about in the nucleus, and are often bent into loops. The significant effects of inversions are due to the fact that normal homologous maternal and paternal chromosomes pair exactly, gene by gene, at meiosis. If an individual is heterozygous for an inversion, the chromosome containing the inverted segment often pairs with its normal counterpart in such a way that a loop is formed at the meiotic prophase (Figure 3-6). If crossing over or exchange of parts between homologous chromosomes occurs

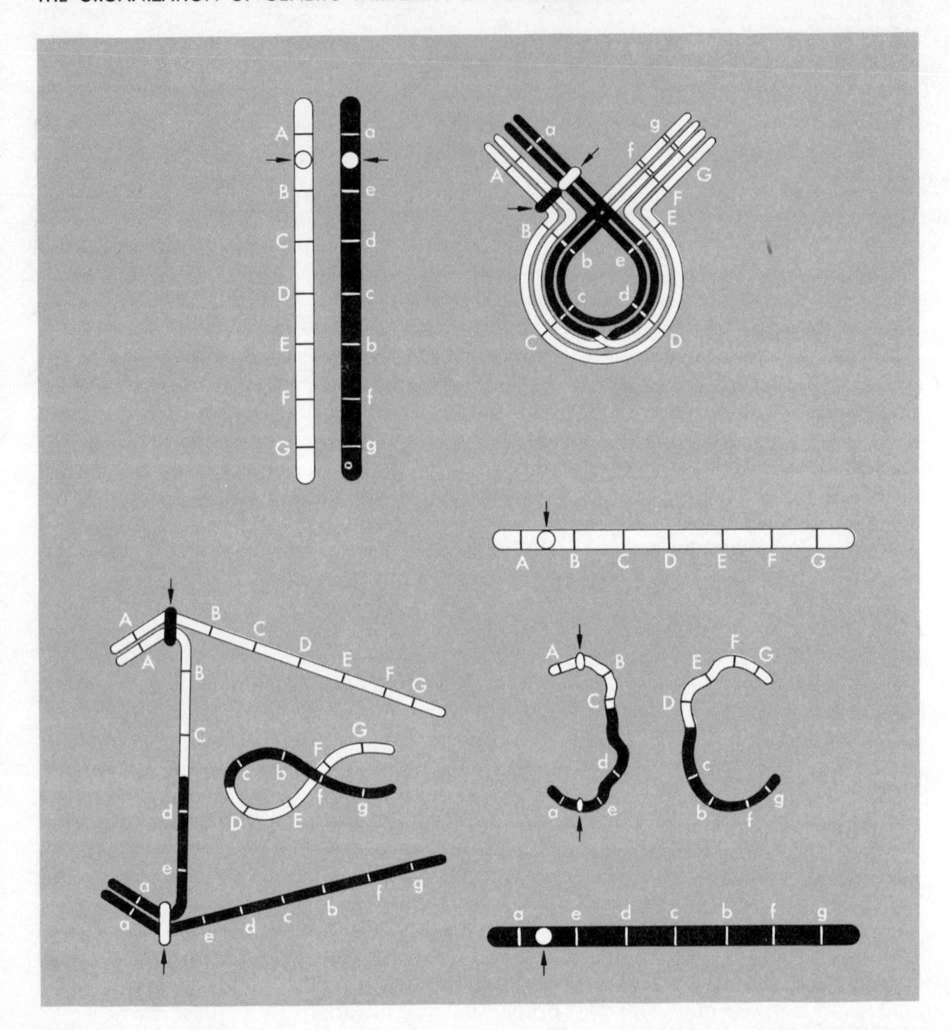

Fig. 3-6. Diagram to show how the effects of crossing over in a segment of a chromosome heterozygous for an inversion can lead to the elimination of chromosome strands containing the recombined genes. From Dobzhansky, *Genetics and the Origin of Species,* 3rd ed. (Columbia University Press).

within the inverted segment, the strands which have exchanged partners form two abnormal structures. One of these is a fragment which lacks a centromere. This fragment becomes lost in the cytoplasm during the meiotic process. The other abnormal strand is attached to both centromeres of the chromosome pair. At anaphase this strand breaks, and its broken fragments, if they ever do enter the cells which result from meiosis, are so lacking in necessary genes that they cause the death of the cell which contains them. In oogenesis of *Drosophila* and probably other animals, these abnormal strands almost always go to the functionless polar bodies, and the egg invariably receives a chromosomal strand in which no crossing over has occurred. If, therefore, an organism is heterozygous for an inversion, the genes in the inverted segment are so tightly linked that they are always transmitted as a single unit. Genetic recombination between them cannot occur.

The fourth type of chromosomal change is known as TRANSLOCATION. Two non-homologous chromosomes may break simultaneously and exchange segments (Figure 3-4). If an organism becomes homozygous for such a rearrangement, some of its genes have been transferred to a completely different chromosome and the linkage relationships of genes are radically altered. This is represented in Figure 3-4 by genes *E*, *F*, *S*, and *T*. Before the translocation occurred, genes *E* and *F* were closely linked and were usually transmitted as a single unit, independent of *S* and *T*. The translocation broke this relationship, and caused *E* and *T* to become closely linked and to be inherited as a unit independent of *S* and *F*. In cases where epistatic interactions affecting adaptiveness occur between such non-allelic genes, the translocation can have a profound effect on the adaptive value of the chromosomes concerned.

In addition, permanent or predominant heterozygosity for translocations may keep together particular combinations of genes located on non-homologous chromosomes and cause them to be inherited as if they were genetically linked. Figure 3-7, shows how this can happen.

Both inversion heterozygotes and translocation heterozygotes exist in natural populations as means of holding together adaptive gene combinations. Heterozygotes for inversions are by far the most common. They have been studied most extensively in *Drosophila* where they are easily recognized in the giant chromosomes of the salivary glands, which have a characteristic pattern of banding (Figure 3-8). In some populations of these flies, such as those of *Drosophila robusta* in the eastern United States and *Drosophila willistoni* in the Amazon Basin of Brazil, nearly every individual is heterozygous for one or more inversions on each one of his chromosomes. That these inverted segments may carry gene combinations of particular adaptive value has been clearly demonstrated both by comparative studies of the distribution of inversion types in natural populations and experiments on competition between inversions under the controlled conditions of population cages. These research studies, carried out by Dr. Theodosius Dobzhansky during the past twenty-five years, provide us with one of the best examples of the adaptive properties conferred by epistatic gene interaction and heterozygosity. He has used a series of inversion types found in *Drosophila pseudoobscura* of western North America. The adaptive properties of these inversion types are clearly shown by their different frequencies of distribution, both in different parts of the southwestern United States which have different climates (Figure 3-9), and at different altitudes in the Sierra Nevadas of California.

This adaptiveness has been demonstrated experimentally in respect to two chromosomal types, Standard (*ST*) and Chiricahua (*CH*), which were obtained from a locality, Piñon Flats, in the San Jacinto Mountains of Southern California. Under natural conditions at this locality the *ST* arrangement is about twice as frequent as the *CH* arrangement during the winter, when little reproduction occurs. During the spring, however, *ST* declines and *CH* increases in frequency, so that by about the first of June *CH* is more frequent

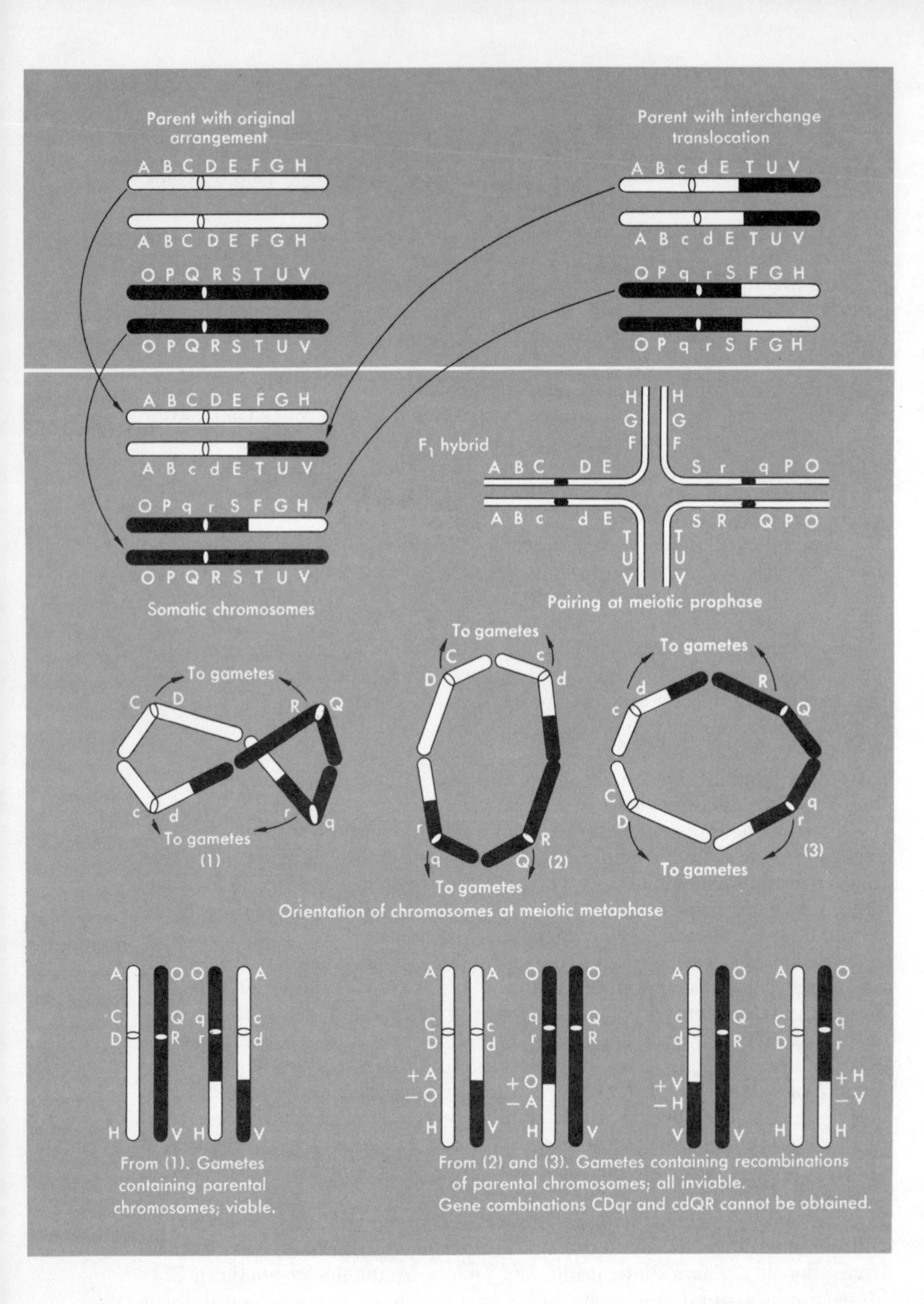

Parent with original arrangement
A B C D E F G H
A B C D E F G H
O P Q R S T U V
O P Q R S T U V
Parent with interchange translocation
A B c d E T U V
A B c d E T U V
O P q r S F G H
O P q r S F G H
A B C D E F G H
A B c d E T U V
O P q r S F G H
O P Q R S T U V
Somatic chromosomes
F1 hybrid
Pairing at meiotic prophase
To gametes
(1)
(2)
(3)
Orientation of chromosomes at meiotic metaphase
From (1). Gametes containing parental chromosomes; viable.
From (2) and (3). Gametes containing recombinations of parental chromosomes; all inviable.
Gene combinations CDqr and cdQR cannot be obtained.

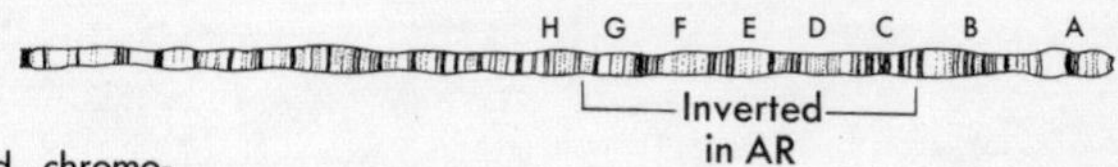

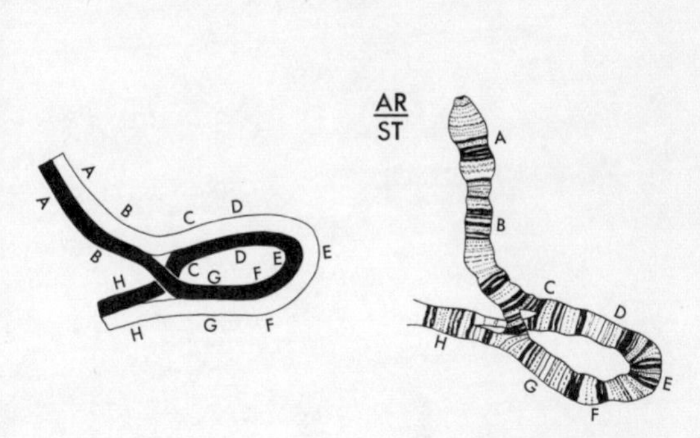

Fig. 3-8. *Top,* actual third chromosome from the salivary gland of *Drosophila pseudoobscura* showing the Standard arrangement and the position of the Arrowhead inversion. *Bottom right,* the loop which is formed in the Standard-Arrowhead heterozygote as a result of gene-by-gene pairing of the normal and the inverted segments of the third chromosome. *Bottom left,* schematic diagram showing how this pairing is interpreted in terms of gene position. From Dobzhansky, *Genetics,* and Carnegie Institute of Washington, Publ. no. 554.

than *ST.* With the onset of warm summer weather, these trends are reversed. By October *ST* has again become twice as common as *CH* (Figure 3-10).

The trend which prevails at Piñon Flats in summer can be duplicated by placing *ST* and *CH* flies in population cages kept at a relatively warm temperature (25°C) and by allowing them to reproduce in uncontrolled fashion so that the cages are crowded with flies (Figure 3-11). Under these conditions, an initially small proportion of *ST* flies will increase in frequency until it reaches 70 per cent and will remain constant thereafter. If the flies are raised in a different fashion, however, other results are obtained. In cages kept at 16.5°C, the relative frequency of chromosomal types in the sample originally introduced remained unaltered for many months; no differential adaptive value could be demonstrated. When bottles were inoculated with a small sample of flies, and by repeated transfers the number of larvae and adults was kept down to fifty per bottle, the *CH* arrangement persisted at a slightly higher frequency than *ST* (Figure 3-11). This duplicates the conditions prevailing during spring at Pinon Flats.

Fig. 3-7. Diagram showing how heterozygosity for an interchange translocation can inhibit recombination between genes located close to the spindle fiber attachment, or centromere, of the chromosomes concerned. *Top row,* two pairs of homologous chromosomes as they exist in related genotypes, one of which contains a reciprocal translocation or interchange. *Second row, left,* the somatic chromosomes of the F_1 hybrid between the chromosomal types represented in the first row. *Right,* chromosome pairing at prophase of meiosis, showing positions of centromeres and of the interchange (center of diagram). *Third row,* the three possible ways in which the ring formed by pairing of the four chromosomes involved in the translocation could become oriented on the spindle at metaphase of meiosis. *Left,* zigzag arrangement which gives rise to gametes of the parental type. *Center* and *right,* open ring arrangements, both of which produce inviable gametes containing large duplications and deficiencies. *Bottom row,* the chromosome constitution of the gametes which would be produced by the three types of orientation shown above. Note that all gametes containing the combinations *CDqr* and *cdQR,* shown at right, are inviable. Hence, this translocation inhibits recombination between alleles at gene loci which are located near the centromeres of nonhomologous chromosomes.

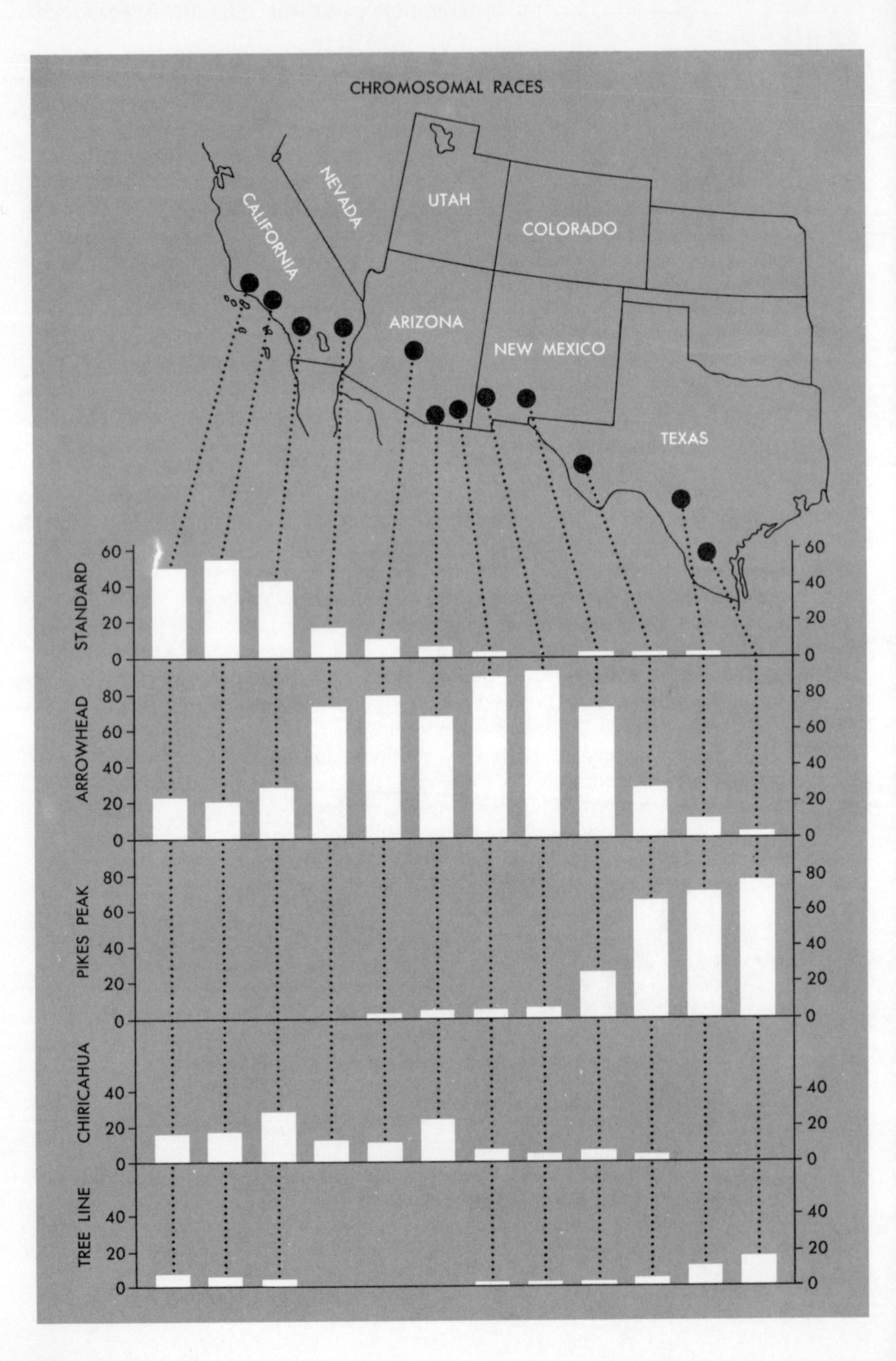
CHROMOSOMAL RACES
NEVADA
CALIFORNIA
UTAH
COLORADO
ARIZONA
NEW MEXICO
TEXAS
STANDARD
ARROWHEAD
PIKES PEAK
CHIRICAHUA
TREE LINE
60
40
20
0
80

Fig. 3-9. Map showing distribution of inversion types in the third chromosome of *Drosophila pseudoobscura* along a transect through the Southwestern United States. From Dobzhansky and Epling, Carnegie Institute of Washington, Publication No. 554, 1944.

In organisms which do not possess giant salivary chromosomes, inversion heterozygosity is much harder to demonstrate than in flies such as *Drosophila*. Nevertheless, studies of chromosome behavior at meiosis in a number of species of plants, such as peonies and rye as well as in some species of grasshoppers, have revealed chromosomal configurations which indicate that they are heterozygous for inversions. There are good reasons for believing that inversion heterozygosity is a widespread device for retaining adaptive gene combinations in populations.

Heterozygosity for translocations, or interchanges between non-homologous chromosomes, is found less often than inversion heterozygosity as a device for retaining gene combinations in populations in the heterozygous conditions. In most organisms the formation of rings or chains of chromosomes at meiosis results in irregular distribution of chromosomes to the gametes, rendering the organism partly sterile. It is, however, widespread in some groups of plants, particularly the *Oenotheras*, "evening primroses," and their relatives. In them, the chromosomes pair only at their ends, and

Fig. 3-10. Diagram showing the change in frequency with the seasons of two inversion types of *Drosophila pseudoobscura* at the Pinon Flats locality in the San Jacinto Mountains, California. From Grant, *The Origin of Adaptation* (Columbia University Press).

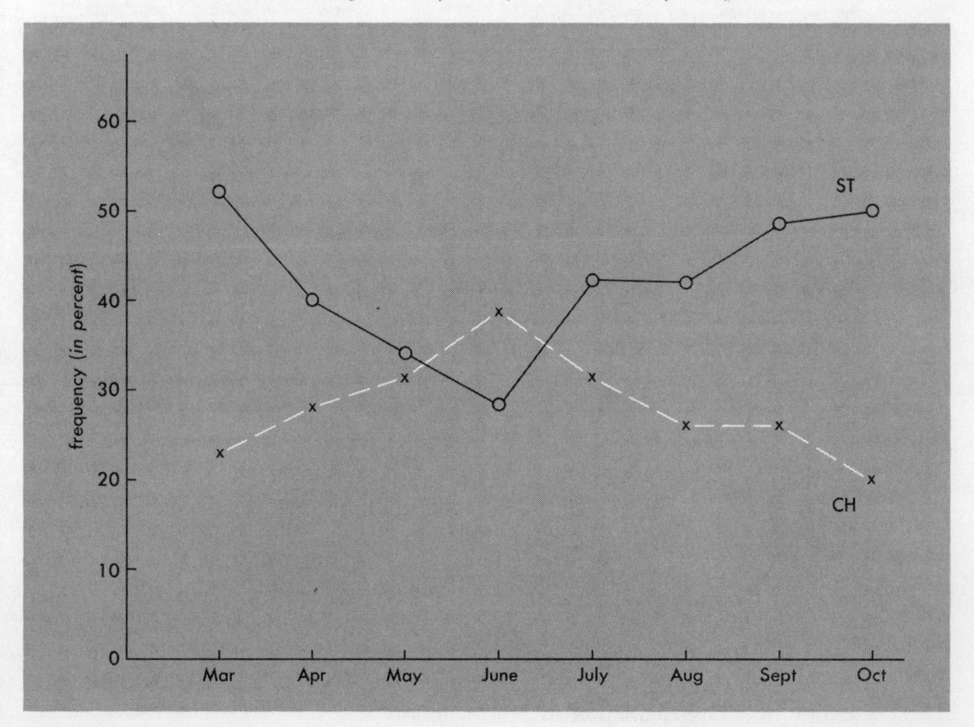

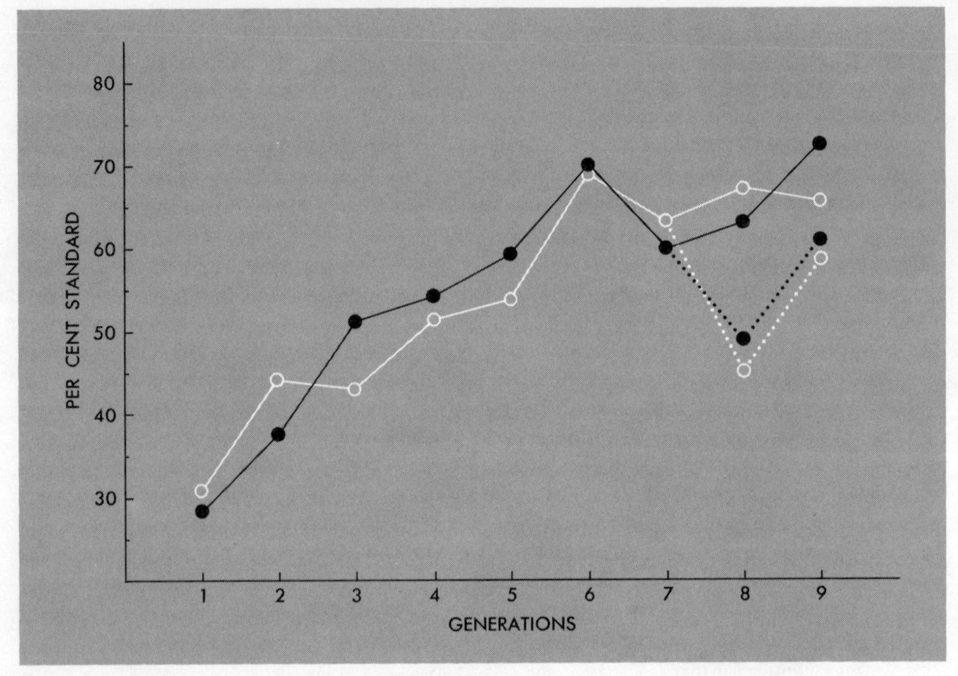

Fig. 3-11. Diagrams showing the results of two experiments involving competition between the two inversion types shown in Figure 3-10. *Left,* at 25°C in cages in which the population of flies was maintained at maximum size and the flies densely crowded, a situation similar to that prevailing in the natural site in middle and late summer. These conditions favor an equilibrium in which the frequency of the Standard arrangement is 70 per cent, and that of

the rings formed in structural heterozygotes are always arranged on the meiotic metaphase spindle in such a way that adjacent chromosomes, which carry homologous segments, pass to opposite poles (Figure 3-12). This results in complete fertility.

Most of the races of *Oenothera* found in eastern North America are heterozygous for a series of segmental interchanges involving all fourteen of their chromosomes, so that at meiosis they form a ring of fourteen rather than seven pairs (Figure 3-12). Since adjacent chromosomes always go to opposite poles, independent assortment of chromosomes is impossible, and the plants can produce only two types of gametes. Meiosis and gamete development have, moreover, been modified in such a way that only one of the two types of gametes can form viable pollen, while the other type is found in megaspores which develop into embryo sacs and produce egg cells. Consequently, a particular race always breeds true for its heterozygous condition. New races can be formed by occasional crossing between preexisting races or by abnormal segregation of chromosomes in the ring. The evening primroses are a remarkable example of a highly evolved and complex chromosomal behavior which achieves the same objective that human plant breeders have achieved with hybrid corn—hybrid vigor combined with constancy of a highly adapted series of genotypes.

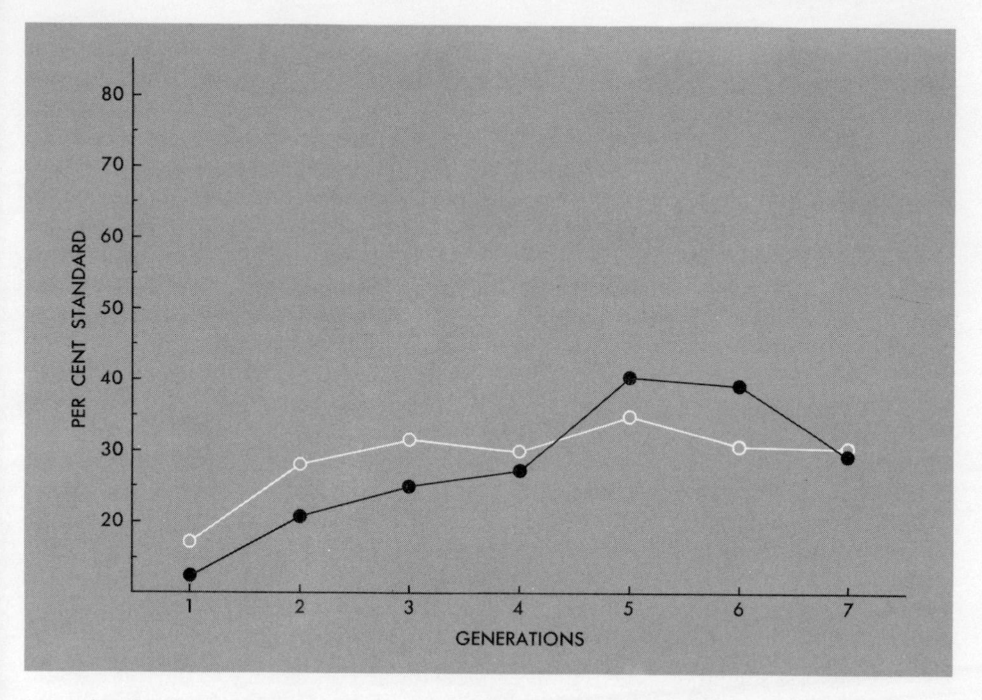

the Chiricahua arrangement is 30 per cent. *Right,* competition between the same chromosomal types in cages kept at 25°C but with flies periodically removed so that their density was always low. These conditions somewhat simulate those prevalent in the natural site in spring and result in an equilibrium at 30–40 per cent Standard and 60–70 per cent Chiricahua. The two lines and two types of circles on each diagram represent two experiments performed under the same conditions. From Birch, Evolution, 9:389. 1955.

Populations which maintain complexes of genes in a balanced condition because of heterozygote superiority cannot be regarded as mere collections of genetically independent individuals. In such populations, individuals which have inferior phenotypes can be valuable to the population as a whole because they can contribute genes which will combine with their opposite alleles to form superior heterozygotes. Under such conditions, the individual no longer forms the unit of selection. Individuals having the same phenotype, and therefore subject to the same selective pressures, may differ radically from each other in respect to the hidden variability which they contain, and consequently in respect to the way in which their death or reduced fecundity will affect the population as a whole. As geneticists have demonstrated with increasing clarity that the gene pool in natural populations is a highly integrated system, evolutionists have shifted their attention from the individual to the population as the unit of natural selection and evolutionary change.

In many natural populations the genes which form a balanced combination are all located close together on a segment of a chromosome, so that they are tightly linked with each other and usually segregate as a single Mendelian unit. Such chromosome segments are called SUPERGENES. They are particularly likely to be found in populations in which two alternative phenotypes for the group of genes concerned both have a high adaptive value in

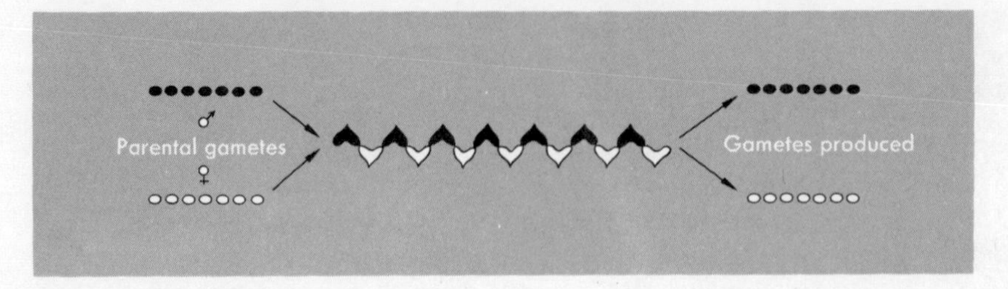

Fig. 3-12. Diagram showing how the system of translocations found in species of "evening primrose" (*Oenothera*) results in the production of only two linkage groups. From Cleland, R. E., *Scientific Monthly.*

relation to each other. A good example is the group of genes which controls the two types of flowers existing in populations of most species of primroses *Primula* and their relatives. This phenomenon, known as HETEROSTYLY, was analyzed by Darwin and has interested many botanists ever since his time. In these plants two types of flowers are found in every population in approximately equal proportions. (Figure 3-13). One of these, known as PIN, has a long style above the ovary, so that the stigma is placed level with the summit of the tube of the corolla. The anthers of its stamens are located in the middle of the corolla tube. In the other type, known as THRUM, the style is short, so that the stigma is situated in the middle of the corolla tube at the same height as are the anthers in the pin type. The anthers of thrum flowers are at the summit of the corolla tube. This placement of anthers and stigma in the two types makes cross pollination easy since a pollinator which alternately visits pin and thrum flowers will receive pollen from one type on the same part of its body which will be closest to the stigma of the other type. Pin and thrum flowers differ from each other in respect to three other characteristics. Their pollen grains are of different sizes, their stigmas are of different shapes and have different kinds of cells or papillae over their surfaces, and the physio-

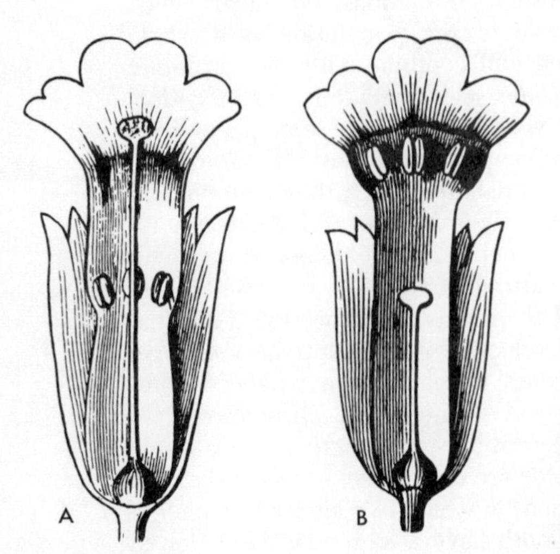

Fig. 3-13. Longitudinal sections through the flowers of a species of primrose (*Primula officinalis*) showing the difference between pin (*left*) and thrum (*right*) genotypes. From Knuth, *Handbuch der Blütenbiologie.*

logical reactions of the stigmas to the growth of the pollen tubes are different. All of these differences promote cross pollination between pin and thrum. The pollen grains of pin fit the papillae of thrum stigmas better than those of their own stigmas. The cells of pin stigmas have an inhibitory reaction against the growth of pollen tubes from pin pollen but promote the growth of tubes from thrum pollen.

Genetic studies have shown that the inheritance of these two flower types is just like that of the two sexes in an animal. One flower type, pin, is homozygous, so the few offspring which can be obtained from self pollination or from pollination with other pin plants always resemble their parents. If thrum plants (*Ss*) are selfed or intercrossed, the few offspring produced include thrum and pin types in a typical 3 : 1 Mendelian ratio. Thrum is, therefore, dominant over pin.

From what we know about gene action, we would not expect to find a single gene which could produce simultaneously abrupt changes in such different characteristics as the length of the style, the position of the anthers, the size of the pollen grains, and the compatibility relationships between pollen tubes and stigmas. It is not surprising, therefore, that by examining many large progenies of thrum × pin crosses in several species of primrose, geneticists have been able to find rare examples of recombination between different components of this character complex. These are best explained by assuming that crossing over has occurred between tightly linked pairs of genes. The genes responsible for these characters are, therefore, situated very close to each other on a single chromosomal segment, so that they are normally inherited as a unit.

We can suggest two hypotheses to explain how such a cluster of genes evolved. One is that the right mutations just happened to occur at a series of adjacent loci, each of which produced one of the characteristics by which the thrum and pin types differ. Once this fortunate combination had appeared its high selective value preserved it. The difficulty of this hypothesis is the unlikelihood that molecules of DNA having the capacity to mutate in the various desired directions would be situated next to each other in the ancestral form. A more likely hypothesis is that the different mutations which contributed to the gene complex took place at various gene loci, and that chromosomal changes such as translocations and inversions brought them together, one by one. An individual in which two of the more important genes had become linked in this fashion would immediately acquire a high selective value, and its valuable combination would quickly spread through the population.

Circumstantial evidence in favor of the second hypothesis has been found by comparing the types of segregation for color pattern found in different species of grouse locusts. In one of these, *Acridium arenosum*, different features of the pattern are controlled by thirteen different pairs of genes which, although situated on the same chromosome, are far enough apart so that they recombine freely by crossing over. In a second species, *Apotettix eurycephalus*, corresponding genes controlling similar features of the pattern are close enough together that they form two tightly linked groups, between which there is about seven per cent crossing over. A third species has many genetic differences controlling the same type of pattern, all of which are inherited as a single unit. This suggests that the evolution of these locusts may have involved progressive restrictions in genetic recombination which have tied together an increasingly large number of genes concerned with a presumably adaptive color pattern.

In a species of land snail found in Europe, *Cepaea nemoralis,* the adaptive value of a supergene has been clearly demonstrated by a long series of experiments conducted by Drs. A. J. Cain and P. M. Sheppard. Populations of this snail are found in a great variety of habitats in Britain; woodlands, meadows, hedges, fencerows, and others. They are all polymorphic for a series of color patterns which involve two elements, the color of the background pigment over the shell as a whole and the presence or absence of bands. Populations found in woodlands consistently contain a higher proportion of pink or brown individuals which lack bands on their shells, while populations from meadows contain a relatively high proportion of yellow, banded individuals. These colors form a concealing camouflage in each habitat, since the background of dead leaves in woodlands is darker and more uniform than is that of the grassy meadows. The value of this camouflage to the snails could be checked by observing that their chief predators are thrushes. These birds take the shells to rocks, known as ANVILS, where they crush them and eat the contents. Among the snail shells found around the anvils, which provide a sure record of the snails which succumbed to predation, the proportion of non-camouflaged individuals was always higher than it was in the surrounding living populations.

Cain and Sheppard found by genetic studies that the depth of color in the BACKGROUND of the shell and the presence or absence of banding are usually inherited as a single Mendelian unit, but that a small percentage of recombination occurs. Background color and banding are, therefore, controlled by different loci which have been united to form a SUPERGENE.

The evolutionary significance of heterozygote superiority in natural populations goes beyond the mere maintenance of adaptive fitness conferred by hybrid vigor. In complex organisms like higher animals and plants, adaptation to new environments requires the establishment of entire new gene complexes involving scores of genes. For instance, if a species of carnivorous animal is to become adapted to a new kind of prey, its limbs must be modified for capturing efficiently this new type of animal, its teeth must be modified for tearing and chewing it, and its digestive system must become modified in adaptation to the new type of food. When a change in the environment takes place, the most successful population will be the one which can most quickly alter its gene pool to take advantage of the change. The occurrence of new mutations at a large number of gene loci, followed by assortment of these mutations into the right combination, is obviously a very slow method of bringing about such a change. Adjustment can be much more rapid if there are some genes among those already existing in the pool which can immediately shift the population toward the new adaptive peak. Consequently, those evolutionary lines which will have the greatest success in the long run will be those having a large store of concealed or partly used genetic variability. Some of this variability can then serve as an initial adaptation to any one of a variety of changes which the environment might undergo. This extra store of variability carried by all populations of cross breeding organisms has been termed its EVOLUTIONARY FLEXIBILITY. We can regard it as a sort of genetic insurance.

Chapter Summary

The existence of hidden genetic variability in cross fertilizing populations has been demonstrated both by inbreeding and by long term selection experiments on both higher animals and plants. Some of this variability exists because of a lag between the occurrence of mutations and their establishment or rejection by natural selection. This is termed the mutational load of genetic variability. In addition, much genetic variability is held in populations because certain gene combinations in the heterozygous condition confer a particular adaptive advantage as compared to homozygotes for the component genes. This advantage can be demonstrated by comparing actual frequencies of genotypes in populations with those expected on the basis of the Hardy-Weinberg law of genetic equilibrium. This is called the balanced load of genetic variability. If two very different groups of genes are held in the population in the balanced state, the condition known as BALANCED POLYMORPHISM results.

Four different kinds of gross alterations of chromosome structure are possible. DELETIONS, in which a segment of a chromosome has been lost, are of relatively little importance in evolution, because of their harmful effects. DUPLICATIONS of a chromosomal segment can promote further differentiation, and can also simulate mutations. INVERSIONS and TRANSLOCATIONS of chromosomal segments, when present in the heterozygous condition, can increase genetic linkage and so bind together adaptive gene combinations. Examples of this phenomenon are described in *Drosophila, Oenothera, Primula,* grouse locusts, and snails. The importance of such increased linkage is due to the number of diverse genes which must contribute to any adaptive mechanism in a higher plant or animal.

Questions for Thought and Discussion

1. Would it be better for populations to consist entirely of homozygous individuals which would breed true for optimum adaptation to their environment, than to possess a store of concealed variability, as most of them apparently do? Give as many reasons as possible for your answer.

2. In many groups of higher plants, particularly cereal grasses like wheat, oats, barley, and their wild relatives, cross fertilization occurs only rarely, and the plants are homozygous at the great majority of their gene loci. Nevertheless, these plants are highly successful both as crop plants and as weeds. Can you provide an explanation for this situation? What might you expect to be the evolutionary future of such populations?

3. Explain the evolutionary importance, if any, of the following chromosomal changes: deletions, duplications, inversions, and translocations.

The differentiation of populations

CHAPTER 4

One of the foremost evolutionists of our time, Sewall Wright, has said that evolution is basically statistical transformation of populations. Based upon the facts and principles set forth in the previous chapters we can amplify Dr. Wright's statement as follows. From the very beginning of life genetic differences between individuals in populations have existed, and in cross fertilizing organisms this variation has been organized into systems of adaptive gene combinations. As environments have changed, populations have either become extinct or have changed their adaptive norms in accordance with the changes of the environment. Environmental changes have been extremely numerous and complex. They have involved alterations of climate, changes in distribution of land and sea, elevation and degradation of mountain systems, and above all, changes in the composition of plant and animal communities. Faced with this continuous, kaleidoscopic shifting of complex mosaics of environments, related populations have often become modified in different ways in response to similar changes in the environment. In addition, many populations have been able to invade new habitats by evolving new adaptive gene combinations. Distantly related or unrelated evolutionary lines inevitably respond to environmental change by evolving in different directions, at least in respect to some of their characteristics. In this way, the enormous diversity of living things which we see all about us has come into being.

As Darwin first realized, and as has now been confirmed by a large number of carefully controlled experiments and observations on a great variety of organisms, the chief agent by which populations have become modified in response to environmental changes is natural selection. All organisms produce far more offspring than can possibly survive, so that some elimination of progeny is inevitable. Given the genetic variation which exists in populations of all crossbreeding organisms, some of these progeny are bound to be better adapted to their environment than others. These favored individuals are more likely to survive and to produce large numbers of offspring than are less fit members of the population. In this way the genes possessed by the better adapted individuals will spread through the population.

To demonstrate that natural selection is chiefly responsible for the differentiation of populations, we must obtain positive answers to the following questions:

1. Can populations be made to respond adaptively to controlled changes in their environment, and are these adaptive responses due to changes in gene frequencies?
2. Can we demonstrate that adaptive changes of this sort have actually taken place in natural populations in response to recorded changes in the environment?
3. To what extent can the differences which naturalists have found between races, species, and higher categories be ascribed to the action of natural selection?

In this chapter, these questions will be answered as well as they can be on the basis of existing information.

The first question has already been answered in part by the experiments of Dobzhansky on changes in frequency of chromosomal types in population cages of *Drosophila,* described in the last chapter. A much larger number of experiments has been performed on artificial cultures of bacteria. Microorganisms are particularly favorable for experiments of this type because very large populations can be raised in a small space, their reproduction is very rapid, and their relatively simple environment can easily be controlled. In the colon bacterium, a favorite subject for such experiments, growing colonies pass through a generation every 35 minutes. This means that in a single day they pass through about the same number of generations that *Drosophila* flies do in ten months, annual plants in 40 years, and human populations in 1000 years. Since the effects of natural selection on evolution are measured in the amount of change per generation, the amount of evolution which can be followed during the normal course of an experiment is obviously much greater in microorganisms than in higher animals and plants.

The fact that populations of microorganisms become adapted to changing environments by selection of favorable mutations has been clearly demonstrated by a number of experiments on the acquisition of resistance to attacks of virus and to the lethal effects of antibiotics. The first experiment was carried out by Luria and Delbruck in 1943. They exposed colonies of colon bacterium to a bacteriophage virus, and observed the frequency of resistant colonies found after exposure. When a series of different samples of the same culture were thus exposed, the variance between samples in the number of resistant colonies (a reflection of the number of resistant bacteria present in the sample at the time of exposure) was about equal to the mean number for the whole series. Mathematical calculations, based upon the Poisson distribution of probabilities, indicate that this amount of variance is expected in any random series of samples from the same population. If, however, different cultures from the same strain were raised for several hours and then sampled, the variance between cultures in frequency of resistant colonies

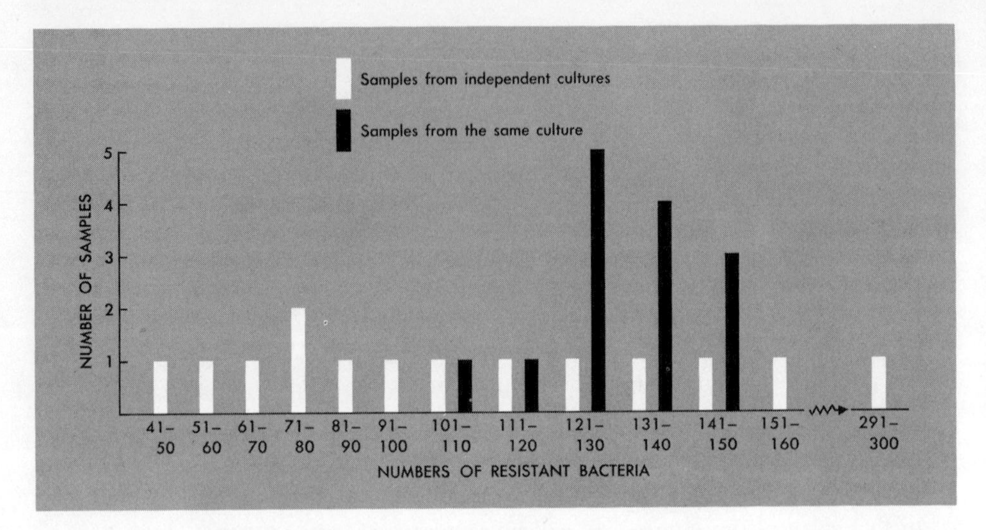

Fig. 4-1. Diagram showing the wide distribution of frequencies of streptomycin resistant bacteria in samples from 15 independent cultures of *Staphylococcus* exposed to the drug, as compared to the much lower variability, and approach to a normal curve, found in 15 samples from a single culture. From data of Demerec, 1948.

was several times as great (Figure 4-1). This showed that the appearance of resistant colonies was due to chance mutation during the period of incubation and growth before sampling. In some cultures, mutation to resistance occurred early in the growth period, so that resistant cells were relatively common at the time of sampling, while in other cultures the late occurrence of mutations resulted in the presence of only a small number of resistant bacteria in the cultures at the time of sampling.

A second method of proving this same fact was carried out by Dr. Joshua Lederberg, as follows. A large number of colonies on a Petri dish, some of which were believed to consist of resistant bacteria, was duplicated by imprinting some individuals from each colony on a second dish. The position of the colonies on the two dishes was kept the same because they were transferred through their adherence to a round piece of velvet which exactly matched the dish (Figure 4-2). One of the twin dishes was then exposed to streptomycin which killed all colonies except those resistant to the drug. Doctor Lederberg then went to the twin dish which had never been exposed to streptomycin, took a few bacteria from the replicates of those colonies which had resisted streptomycin, as well as those which had succumbed, and placed them in test tubes filled with a medium containing streptomycin. In every case he found that colonies were equally resistant both in the replicate which had been exposed to streptomycin and in the one which had not, while none of the bacteria from replicates of colonies which had been killed by streptomycin acquired resistance when placed in test tubes containing the drug. This showed that the resistant colonies had grown from a single resistant bacterial cell, which had mutated to resistance before the streptomycin was applied. The alternative explanation, that resistance to streptomycin is acquired by exposure to the drug, and that this acquired resistance is then inherited, has been completely ruled out by Dr. Lederberg's experiment.

Another experiment using a different antibiotic has shown that bacterial populations can acquire resistance to very high doses of antibiotics by a succession of mutations. In an experiment on the resistance of colon bacteria to puromycin, L. L. Cavalli and G. A. Maccacaro were able to show that resistance to a concentration of the drug 250 times as great as that tolerated by normal bacteria could be acquired by a succession of transfers of the cultures to gradually increased concentrations (Figure 4-3). Since the strain

Fig. 4-2. Diagrams illustrating the duplicate plating technique of Lederberg, to demonstrate that streptomycin resistance results from mutations which can occur quite independently of exposure of the culture to the drug. Explanation in the text. From Sager and Ryan, *Cell Heredity* (John Wiley & Sons, Inc.).

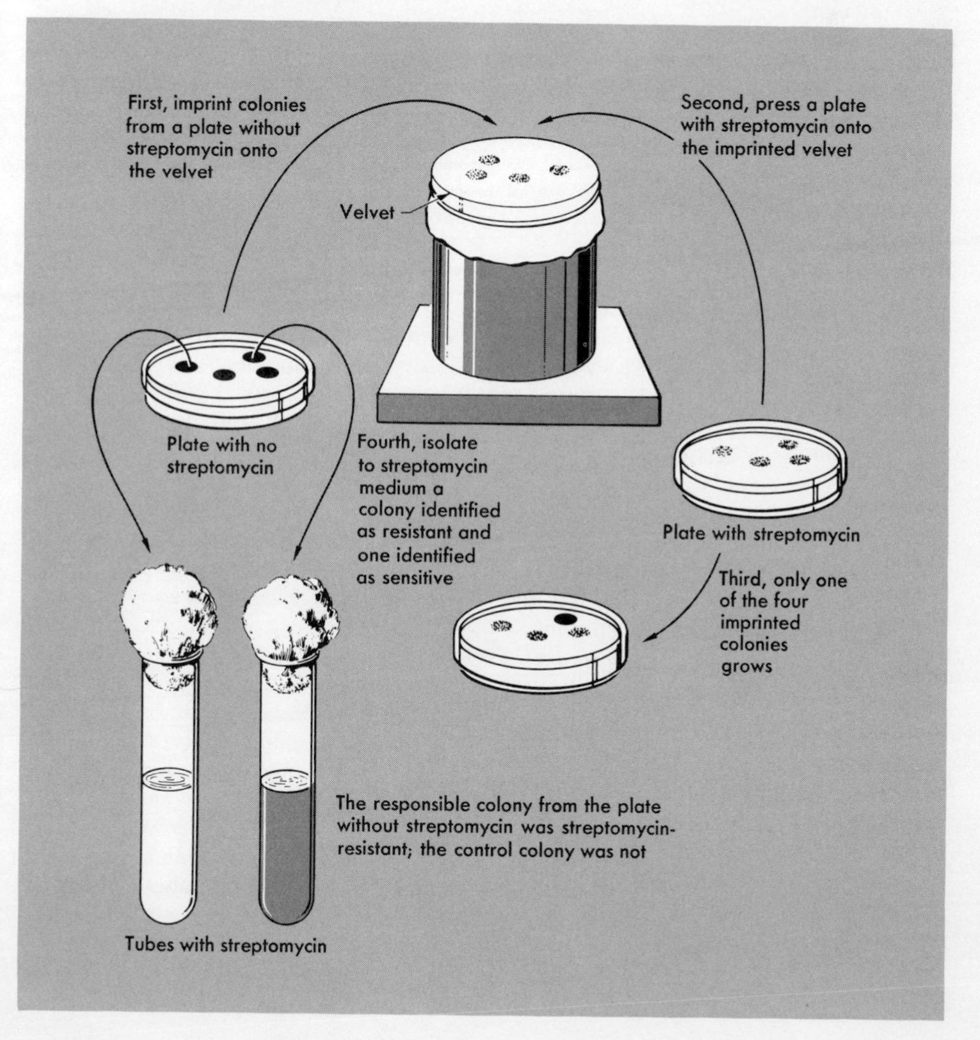

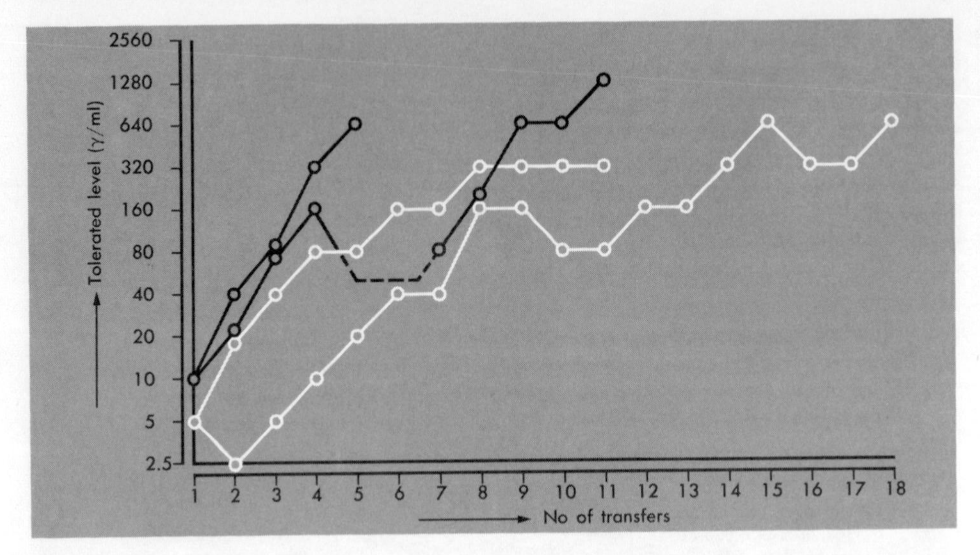

Fig. 4-3. Diagram showing the results of four separate selection experiments which illustrate the stepwise acquisition of high resistance to the antibiotic chloramphenicol, acquired by exposing cultures of *Escherichia coli* to gradually increasing concentrations of the drug and selection of mutations to successively greater resistance. From Cavalli and Maccacaro, Heredity, 6:311. 1952.

of bacteria used reproduces sexually, the resistant bacteria could be crossed to descendants of their susceptible ancestors. If strains with intermediate resistance were used in such crosses, simple Mendelian inheritance was found. Highly resistant strains crossed with susceptible ones gave complex segregations, indicating that resistance had been acquired by mutations at several different loci. Since strains marked with previously analyzed genes were used, linkage was found with known gene loci so that the position on the bacterial chromosome of some of the genes for resistance was determined. Puromycin resistant strains grow much more slowly than susceptible ones, so that when the resistant cultures were again raised on puromycin-free media, resistance was lost through the occurrence and establishment of reverse mutations toward susceptibility.

The experiments described above enable us to answer "yes" to the first question posed at the beginning of this chapter. A positive answer to the second question—that adaptive changes actually have taken place in natural populations due to the action of natural selection—is much harder to obtain because the experimenter cannot set up his own conditions. He must find situations in which he suspects that such changes have been taking place, and in which historical records are available to give him some idea of what populations were like in past years. The best analyzed example of this type is the change in color which has taken place in certain populations of moths in the industrial regions of Europe during the past hundred years. This phenomenon, an increase in frequency of dark colored mutants, is known as INDUSTRIAL MELANISM. Populations of several different and unrelated species of moths have changed color over essentially the same period of time. Pre-

Fig. 4-4. Photographs showing the inconspicuousness of the normal form and the conspicuous appearance of the melanic form of the peppered moth (*Biston betularia*) on a normal, lichen covered tree trunk and the reverse situation when the same forms are on soot-darkened, lichen-free bark. Photographs by S. Beaufoy, from Wallace and Srb, *Adaptation* (Prentice-Hall, Inc.).

vious to the change, the moths of these species were uniformly pale gray or whitish in color. At various times and places, which varied with the species concerned, individuals with dark colored wings and body appeared, first as isolated freaks, and later with increasing frequency until they dominated the populations of certain areas. These were in every case regions of extreme industrialization, such as the Ruhr Valley of Germany and the English Midlands, where the smoke and soot from factory chimneys darkened the ground and the tree trunks (Figure 4-4).

Genetic tests showed that the dark individuals which appeared were in each case dominant mutations, which produced typical Mendelian segregation in crosses with normal pale forms of the same species. Direct experimental tests, as well as the discovery of similar dark mutants as extremely rare aberrations in regions far from industrialization, showed that the mutations were not induced by the heavy metals and other impurities added to the diet of the caterpillars which had to feed on soot-covered leaves. The effect of industrialization was to increase the frequency of black moths, which had always been present as rare spontaneous mutations, but in non-industrial areas are quickly eliminated from the populations by adverse selection.

Professor H. B. D. Kettlewell of Oxford University set out a few years ago to demonstrate experimentally the basis of this change. He noticed that in the woodlands of southwestern England, far from the effects of industrialization, the normal pale colored moths, resting upon the pale, lichen-covered bark of the tree trunks, are perfectly camouflaged, while the dark mutations, resting on these same tree trunks, are very conspicuous. On the other hand, the dark bare trunks of the trees in woodlands near the highly industrialized city of Birmingham exactly match the color of the dark mutants of the moths, while the normal pale forms are very conspicuous when resting on them. He therefore suspected that birds, which are the natural predators of these moths, would easily find and destroy the dark mutants when resting on the pale tree trunks of non-industrial areas, and would similarly overlook the dark mutants and feed on the pale moths in the industrial areas. In order to test his hypothesis he raised thousands of moths belonging to the peppered moth species, and released them in two separate experiments. Having previously marked each moth with a little spot of paint, he could, later on, trap moths in the same area, and distinguish between individuals native to the area and those which he had previously released. In the non-industrial area of Dorset, he later recaptured 14.6 per cent of the pale moths released, but only 4.7 per cent of the dark moths. In the industrial area near Birmingham, the situation was reversed; he recaptured only 13 per cent of the pale moths, as compared to 27.5 per cent of the dark mutants. To demonstrate that selective destruction by birds was the reason for this difference, he set up blinds, and with companions watched the birds attack moths which he placed upon tree trunks. These observations, supported by excellent motion pictures which are available for anyone to see, showed clearly that birds destroy many more dark mutants than pale moths in the non-industrial areas and more pale than dark moths in the vicinity of Birmingham. The action of natural selection in producing this small but highly significant step of evolution is, therefore, clearly documented.

Some other examples exist of historically documented changes in the gene pools of populations which have been brought about by selection of adaptive mutants or gene combinations. One is the increased resistance to artificially applied sprays on the part of scale insects infesting the citrus orchards of Southern California. Another is the acquisition of resistance to DDT which has taken place in populations of house flies throughout the United States during the past twenty years since this chemical has become commonly used to combat the pest. All of these examples indicate that as research data on the history of natural populations increase in quantity, the answer to the second question will become increasingly positive.

THE NATURE OF COMPETITION

The experiments which have just been described tell us a lot about the ways in which organisms compete with each other, and how one population can gain an evolutionary advantage over another. The most important lesson to be learned from them is that physical combat, which results in the death of the less successful individuals, is one of the least common ways in which the "struggle for existence" takes place. It was not a factor in any of the examples described above. In all of the experiments with bacterial populations, the successful genotypes were those which could best resist a drastic change in the environment, namely the addition of disease causing viruses or toxic antibiotics. In the example of industrial melanism, too, the successful genotypes were those which, under the new conditions of industrialism, could best escape from their normal predators. Consequently, success in avoiding or adjusting to unfavorable factors of the environment must be regarded as one of the most important factors of competition which leads to natural selection.

Another type of competition which very commonly leads to natural selection is the superior ability of an organism to take advantage of its surrounding medium, so that it grows faster and crowds out its less successful competitors. This kind of competition is very common in plants. If not removed from a field, weeds will grow faster than the crop which has been planted, and, although they rarely kill the crop plants outright, they often reduce or eliminate their ability to make seeds and reproduce. In many of the warmer parts of the United States, Bermuda grass (*Cynodon dactylon*) will establish itself in lawns. Because it can make more efficient use of the soil, water, and temperature regime available to it, this weedy grass will often crowd out and actually kill the more desirable but less tough lawn grass. Among animals this type of competition has been recorded in sedentary forms, such as barnacles.

Natural selection can also exert its effects through the better ability of the favored genotype to reproduce and leave a larger number of viable offspring, even though all of the individuals of less successful genotypes grow to maturity and actually produce some progeny. The action of this factor of selection was clearly shown in a series of experiments carried out by Dr. Joshua Lee on competition between varieties of cultivated barley. Previous data had shown that when a mixture of four varieties, each of them constituting one fourth of the mixture, were sown in the field, differences in the number of seeds which each plant produced caused the proportions of the four varieties in the seeds harvested to be very unequal. If a small sample of this harvest was planted in the same field the following year, the second harvest had a still greater proportion of the more successful variety and still less of the unsuccessful one. When this procedure was repeated for several generations, the less successful varieties were completely eliminated from the mixture.

Doctor Lee studied the behavior of two of these varieties when plants were sown in a field at known densities of spacing, and alternating with each other in a regular fashion. By keeping track of each plant he found that all plants grew to maturity and produced some seed. Nevertheless, the amount of seed per plant produced by one variety was sufficiently greater than that produced by the other so that the different proportions of seed found in the earlier experiment could be explained. In this way, Lee showed that one genotype can replace another in a population, even though all individuals which establish themselves as seedlings can grow to maturity and produce some seed. The selective differential lies not in the greater probability that individuals of the less successful genotype will die, but in the smaller number of seeds which they produce. If seeds are destroyed at random, which would happen in nature through the attacks of seed eating animals or through the deposition of seeds in places where the seedlings are unable to grow, the success of the more fecund genotype producing the most seeds would be the direct result of its greater seed production. By this means, it would contribute a larger number of seeds to the surviving sample.

In animals the same type of selection factor would operate in populations of rapidly reproducing organisms which have no particular mechanism for defense against predators. The small crustaceans and protozoa which form the floating plankton in the sea are good examples. If the young offspring of a species are destroyed indiscriminately by a great variety of predators, evolutionary success is best assured by producing a large number of zygotes and young embryos. Organisms which have evolved better means for caring for their young are, however, better adapted if they produce a smaller number of zygotes and young embryos.

THE ORIGIN OF COMPLEX ADAPTIVE SYSTEMS

The statement was made at the beginning of this book and has been repeated directly or by implication throughout the discussion, that natural selection is the principal and perhaps the only significant guiding force in evolution. If this is so, then all differences between populations, including races, species, and higher categories, should be either directly or indirectly connected with different systems of adaptation. Consequently, one of the major tasks of the evolutionist is to show how such systems can arise, and how they can be responsible for the differences between the various kinds of animals and plants which we see around us.

A short book, such as this one, permits a discussion of only a small part of the extensive research which has been carried out in recent years on the evolution of adaptation and adaptive differences between organisms. For fuller accounts the reader is referred to a companion volume to this one, *Adaptation*, by Wallace and Srb, as well as to the longer books by Ford, Grant, and Mayr, cited at the end of this volume. The present discussion will center about answers to two questions which are often asked:

1. How can natural selection guide populations into the complex, elaborate, and specialized adaptive systems which we often find in nature?
2. How can natural selection control the differentiation of characteristics of which the adaptive significance is not easily recognized, and which appear to be non-adaptive?

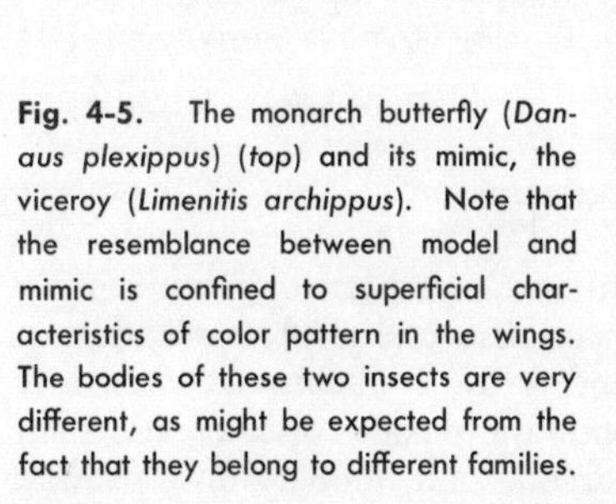

Fig. 4-5. The monarch butterfly (*Danaus plexippus*) (*top*) and its mimic, the viceroy (*Limenitis archippus*). Note that the resemblance between model and mimic is confined to superficial characteristics of color pattern in the wings. The bodies of these two insects are very different, as might be expected from the fact that they belong to different families.

A group of elaborate and specialized adaptive systems which have been so well analyzed by both observation and experiment that their mode of evolution is reasonably well understood are those involving mimicry. This phenomenon is known through observations which naturalists have made on various kinds of animals for more than a hundred years. Most insects are palatable to birds and other predators, from which they are protected by their inconspicuous colors. These colors blend into the surrounding background and provide camouflage. On the other hand, some species of insects are distasteful, either because they exude smelly or bitter secretions or because they possess powerful stings. Many of these species are not camouflaged, but on the contrary are conspicuously colored so that they are easily seen by birds and other potential predators. This type of conspicuous coloration of a noxious animal is known as WARNING COLORATION. Good examples are skunks, with their conspicuous black and white bands, as well as bees and wasps, conspicuously striped with yellow and black. In many instances, animals with warning coloration are imitated by completely unrelated species which are not noxious and are perfectly palatable to birds and other predators. For instance, bees are imitated or mimicked by a group of flies, known as bee-flies (Asilidae), which resemble them both in superficial features of their color pattern and in their behavior (Figure 4-6). In the eastern United States the large orange and black butterfly known as the monarch, (*Danaus plexippus*), a noxious species, is mimicked by the palatable viceroy (*Limenitis archippus*), a butterfly belonging to a completely different family (Figure 4-5). The resemblance of mimic to model is merely superficial. In basic features of body structure, mimics resemble their very differently colored relatives. They

Fig. 4-6. A bumble bee (*Bombus*) and its robber fly mimic, *Dasylis socrata*.

are like their unrelated models only in the general outline and color pattern of their wings.

Mimicry presents the evolutionist with two problems. First, are we correct in assuming that predators will learn to associate unpleasant odors or stings with conspicuous patterns of color and form? Will this cause them to avoid animals with these characteristics, both the noxious species and their harmless and palatable mimics? Second, assuming that mimicry does provide a selective advantage, how can these superficial but often very striking resemblances arise by means of mutation, gene recombination, and selection?

The first question has been answered in the affirmative by means of a brilliant series of experiments conducted by a husband and wife team, L. P. and J. V. Z. Brower. They showed that two very different kinds of animals, toads and birds, can be conditioned to avoid both warningly colored animals and their harmless mimics. In experiments with captive toads they first placed bees in the cage. The toads captured the bees with their tongues, were stung in the mouth, and showed obvious signs of discomfort. After this had happened a few times, the toads not only avoided bees, but crouched motionless in a corner of their cage whenever these insects were placed into it. They likewise avoided harmless bee-flies. On the other hand, when dragon flies (*Pachydiplax*) were introduced into the cages, these same toads readily ate them. Control toads, which had never been exposed to bee stings, readily ate both bee flies and bees from which the stings had been removed.

The effectiveness of the mimicry possessed by the viceroy butterfly was tested by means of experiments with Florida scrub jays, (*Cyanocitta coerulescens*). Young jays were fed first with viceroy butterflies, which they ate in every test. They were then presented with monarch butterflies, which they tried to eat, but rejected violently after the first taste, and thereafter avoided consistently. After having been presented with fifty monarch models, the jays were again given viceroys, which they now avoided with equal consistency.

The Browers continued their tests, using two other mimic butterflies and their models with equal success. One of these mimics was a close relative of the viceroy, which mimics a species related to the monarch. They made the significant observation that among several birds which had been conditioned to avoid monarch butterflies and their viceroy mimics, three also avoided the related but different mimics. This shows that even an imperfect form of mimicry can give an animal some degree of protection. This fact is very important in connection with the origin of mimicry.

Mrs. Brower then reasoned that if natural mimics are highly effective in deterring potential predators, equally effective mimics could be produced by artificial means. To determine this, she used captive starlings and meal worms, which are their normal food when in captivity. In order to make distasteful "models," she dipped meal worms into a bitter substance, quinine dihydrochloride, and at the same time colored them with a band of green cellulose paint. She made "mimics" by coloring meal worms with the same type of green paint, and dipping them in distilled water. Non-mimetic, edible, painted meal worms were made by using orange paint of the same type and distilled water.

The nine starlings used were each given 160 trials over a period of 16 days. In each trial, a bird was given an orange banded, edible and a green banded meal worm, either a model or a mimic. Different birds were given either 10, 30, 60, or 90 per cent mimics. The starlings ate all of the orange banded meal worms, showing that the paint itself had no effect on palatability, and also that these birds have no instinctive aversion to brightly and conspicuously colored prey. They quickly rejected the bitter, green-banded models, and after having tasted them, avoided the similarly colored "mimics." Eighty per cent of the palatable green-banded "mimics" escaped predation in this manner.

This series of experiments has clearly shown that warning coloration and mimicry are both highly effective adaptations to escape from predation. The probable manner in which such patterns of color and form evolve has been inferred on the basis of a series of observations and experiments by Drs. C. A. Clarke and P. M. Sheppard on an African species of swallowtail butterfly, *Papilio dardanus*. This species has evolved one of the most remarkable series of mimicking races of any species of animal (Figure 4-7). Throughout its range, which includes all of tropical and south Africa as well as the island of Madagascar, the males of *P. dardanus* have yellow and black wings bearing "tails," much like the swallowtail butterflies found throughout the United States. In Madagascar and parts of Ethiopia the females resemble the males. In these regions noxious butterflies of the family Danaidae do not occur, and mimicry has not evolved. Elsewhere, throughout tropical Africa, females of *P. dardanus* are very different from males. Their wings lack tails, and exist in a variety of color patterns. Six of the patterns are mimics of six different species of noxious butterflies belonging to three different genera in two separate families. In most of this vast area which includes more than half of the African continent, populations usually contain two or three different forms of mimicking females, although one form is often predominant. In the East African mountains of Tanganyika and Kenya, however, females of *P. dardanus* occur in a slightly different series of forms which are imperfect mimics of the noxious Danaidae. This is associated with the absence or rarity of these models. The fact that protective mimicry occurs only in females is associated with the adaptive value of protection during the period when the female is producing and laying her eggs.

Fig. 4-7. Non-mimicking and mimicking *Papilio dardanus* with their models. *Left,* Female (*top*) and male (*bottom*) of *Papilio dardanus* from Madagascar, a non-mimicking race. *Top row, Second,* a noxious model, *Danaus chrysippus,* from East and South Africa; *next,* the female of *Papilio dardanus* which mimics it; *right,* an imperfect mimic which was a segregate from a hybrid between the mimicking race and a race not possessing this mimic. *Bottom row*

Crosses between mimicking and non-mimicking races have shown that the principal differences between them are controlled by two independently segregating groups of alleles. One pair of these determines the presence versus the absence of "tails" and the other, a series of multiple alleles, the major features of the color pattern. The most significant fact about the inheritance of these genes is that if hybrids are made between forms inhabiting regions which differ from each other with respect to the presence versus the absence or frequency of the various models, the results are quite different from those resulting from crossing different forms from the same or similar regions. If the two parents both come from a single population which contains two or more different mimics among its females, their progeny are likely to segregate for these different mimicking forms. In this case, the segregation of the progeny is clear cut and complete dominance exists. Consequently only typical, perfect mimics of one or the other "model" species appear among the F_2 or backcross females. If, however, crosses are made between forms inhabiting different regions, which have never been in genetic contact with each other, the F_1 hybrid is usually intermediate. The variation between different F_1 individuals is considerable, while the F_2 or backcross progeny do not show clear-cut phenotypic segregations.

(starting with second figure), A model belonging to another genus (*Amauris niavius dominicanus*, its mimicking female of *P. dardanus*, and an imperfect mimic which segregated from a hybrid with a non-mimicking race. Imperfect mimics are produced by the action of modifier complexes of genes which occur in regions where models are infrequent. From Sheppard, Cold Spring Harbor Symposia, 24:1959.

The only possible explanation for these results is that populations of *Papilio dardanus* contain large numbers of modifying genes at different loci from those of the two major or "switch" genes. The modifying genes affect the expression of the latter in various ways. These form a "genetic background" which has been molded by natural selection to give the maximum proportion of mimicking females in regions where models are abundant. However, the genetic background is variable and indeterminate in regions where models are rare or absent.

According to R. A. Fisher, whose theory has been rendered highly plausible by the observations and experiments described above, the evolution of this system of gene controlled patterns of mimicry was probably as described below. In the remote past both male and female individuals of *Papilio dardanus,* throughout its range, were tailed and had the typical swallowtail color pattern of yellow and black. This hypothesis offers the best explanation of the similarity between non-mimicking populations in Ethiopia and Madagascar at separate margins of the distribution of the species. Then, in some part of central Africa where a noxious species of the genus *Amauris* of the family Danaidae was abundant, recessive mutations for a changed color pattern took place, and either through selection in favor of heterozygous individuals or through inbreeding or both, individuals homozygous for a particular recessive allele appeared. These butterflies bore a general resemblance to a noxious species of *Amauris* and consequently were attacked less often by predators than were their normal relatives. By producing a larger number of offspring, they increased the frequency of the mutant condition until a large proportion of the females were of the imperfectly mimicking form. Once this condition had arrived other mutations which tended to perfect the mimicry immediately acquired a high selective value. The mutation for the tailless condition in females took place in one of these early mimicking populations and spread throughout the region where models occur. Other mutations with smaller effects became established in different local populations and brought about the regional differences in genetic background which were demonstrated by the hybridizations made by Clarke and Sheppard. Once mimicry of the *Amauris* species had become well established and perfected, the occurrence of new mutations of the original "switch" gene, followed by selection of appropriate modifiers, evolved the mimicry of other noxious models.

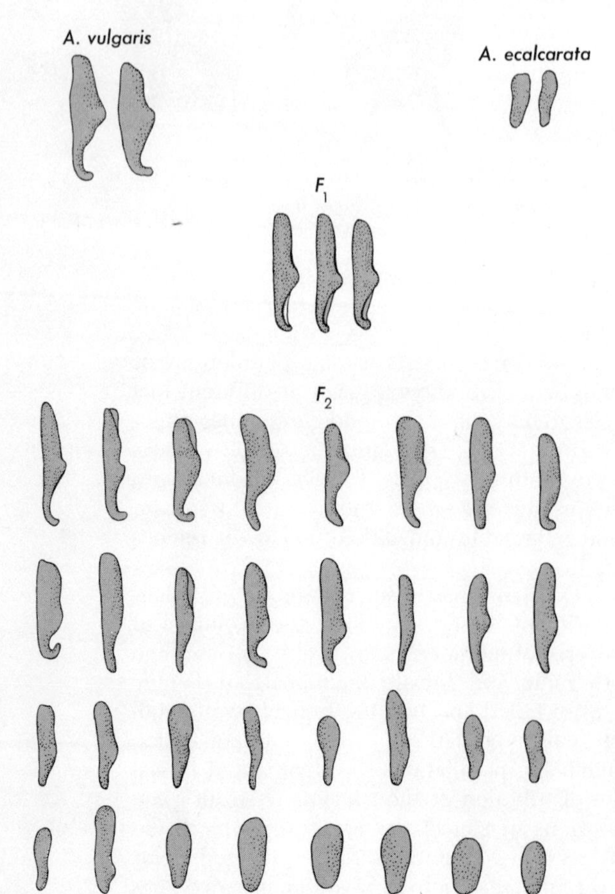

Fig. 4-8. Spur petals of the European columbine, the Asiatic spurless columbine, and of F_1 and segregating F_2 hybrids between them. From Prazmo, *Acta Soc. Bot. Poloniae.*

The pattern of evolution which has emerged from these studies of mimicry, consisting of the initial establishment of mutations in one or two major genes, followed by modifiers which perfect the new adaptive system, has probably occurred many times in various groups of organisms. In higher plants, a probable example is the columbine (*Aquilegia*). This genus of the buttercup family is characterized by flowers having long spurs on their petals. This adapts them to cross pollination by specialized pollen vectors, either bumble bees, hummingbirds, or hawk moths. One species of columbine from eastern Asia lacks spurs altogether. When this species is crossed with spur bearing species, the segregation in the F_2 generation shows that presence versus absence of spurs is governed by a single pair of "switch" genes (Figure 4–8). An additional pair of major genes produces the long spurs of species pollinated by hawk moths. An additional pair of genes determines the difference between bent spurs, an adaptation to pollination by bumble bees, and straight spurs, found in species pollinated by hummingbirds or hawk moths. On the other hand, the length of the spurs, the size of the flowers, and their various shades of color are all determined by many pairs of modifying genes. This situation is best explained by assuming that the mutation for spurs on the petals first transformed the ancestral, anemone-like flower into one which was visited preferentially by bumble bees. Successive changes in the other major genes and the various modifier complexes evolved the numerous variations of the basic and initial flower pattern which gave rise to the various species of columbine that exist today.

THE EVOLUTION OF APPARENTLY NON-ADAPTIVE DIFFERENCES

A variety of answers can be given to the question regarding the origin by natural selection of apparently non-adaptive differences. The first is that many differences are, in fact, adaptive, but their adaptive nature is not apparent to us. An example is red or purple coloring in bulbs of onions. A priori, one would say that there is no particular value to an onion in being either white or red. Yet under certain circumstances the difference is of great importance, since colored bulbs are resistant to attacks of the smudge fungus, *Colletotrichum*, a damaging parasite. This is because they contain catechol and protocatechuic acid, substances which are poisonous to spores of the fungus. A single recessive gene in the homozygous condition produces the color and the fungicidal substances at the same time.

In animals, the bristles or whiskers on the face of the house mouse provide us with another example. At first sight, one would not expect differences in the number of bristles above the eye to have much effect on the viability or evolutionary success of a mouse. Yet this number is remarkably constant in the species, and strains which have been selected for deviating numbers are subnormal in viability. The value of the bristles to the mouse is revealed when mice from which the bristles have been cut are made to run through a maze. They do this much less successfully than normal mice, suggesting that the whiskers over their eyes help mice to find obscure passageways in the dark.

Numerous other examples could be cited. The examples increase in number as one studies organisms increasingly unfamiliar to and different from the human species. A commonplace dictum of modern studies of evolutionary adaptation is that the adaptive superiority of one genotype over another through greater survival and reproduction under controlled conditions is far easier to demonstrate experimentally than are the reasons for this superiority. Consequently, if we cannot see why a difference should be connected with differential adaptation, we are not justified in concluding from this fact that the difference is non-adaptive.

Differences which are non-adaptive as phenotypic characters may arise because they are produced by genes having several pleiotropic effects, some of which are invisible but are strongly concerned with adaptation. A classic example is sickling in human populations. This condition, which consists of a drastic change in the appearance of the red blood corpuscles (Figure 4-9), is produced by the heterozygous condition for a gene (*Ss*) which is lethal in the homozygous condition (*ss*). People having the trait (in the heterozygous condition) are mildly anemic, and so would seem to be at a slight disadvantage. Nevertheless this trait is common in many Negro populations of Africa, and may reach a frequency of 40 per cent. This is because heterozygotes for the sickling gene are more resistant to malaria than are homozygous normal (*SS*) individuals. Consequently, the morphological difference in the appearance of the red blood cells is incidental to a much more important physiological difference, resistance versus susceptibility to malaria. We might expect to find differences in respect to sickling between populations in which malaria is present and those which are free of the disease. Such differences have been found.

Another way in which differences with little or no adaptive significance can become established is through genetic linkage with adaptive differences. No clear-cut examples of such linkage are yet available, although the studies of A. J. Cain and P. M. Sheppard on variations in color pattern in the land snails of Britain provide strong circumstantial evidence for their existence. A

Fig. 4-9. Human red blood cells from normal and homozygous sickle cell genotypes. From McKusick, *Human Genetics* (Prentice-Hall, Inc.).

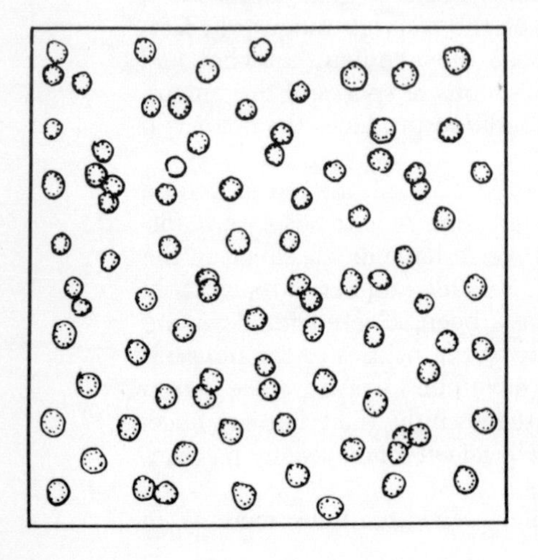

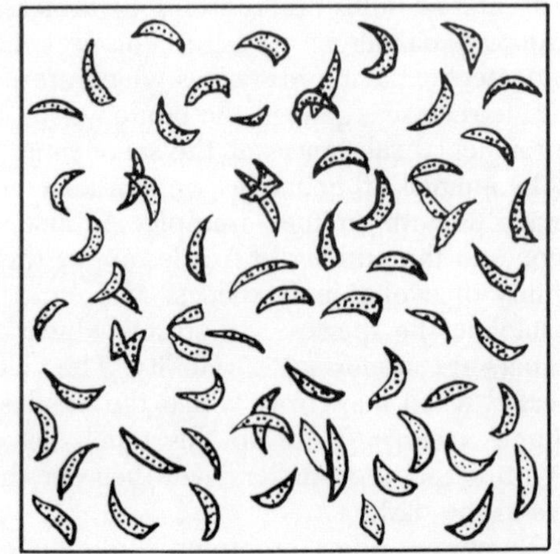

good account of this variation is given by Professor E. B. Ford in his book, *Ecological Genetics*. In one experiment on the fly, *Drosophila melanogaster*, selection for increased numbers of bristles on the abdomen was successful for twenty generations. The flies then became so sterile that selection had to be relaxed, and the bristle number of the flies dropped sharply. When selection was again resumed, increased numbers of bristles were obtained without the accompanying sterility. This result could be explained only on the assumption that during the period when selection was relaxed, linkages between genes for fertility and others for bristle number were broken in some individuals by crossing over. The second period of selection was not accompanied by sterility because it was based upon these individuals.

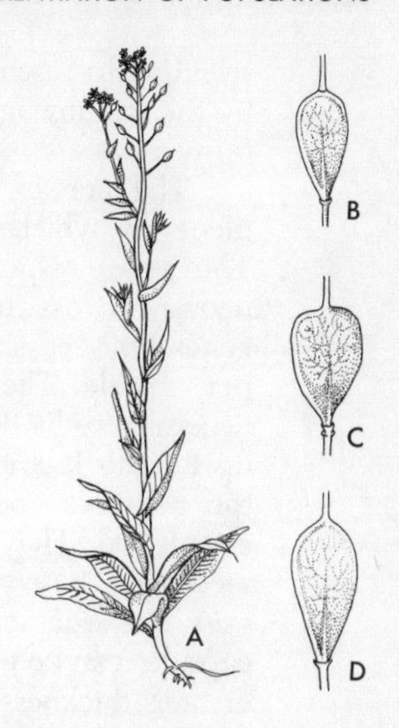

Fig. 4-10. The roadside weed *Camelina sativa* (a and b), together with two types of enlarged capsules (c and d) found in races of the species which infest flax fields. Modified from Hegi, *Illustrierte Flora von Mitteleuropa*, with capsule drawings added from data of Tedin in Hereditas, 6:275.

Genetic differences between populations may arise not only as adaptations to different environments, but also as different ways of adjusting the organism to similar environmental changes. Given the genetic and physiological complexity of any population of higher organisms, many different ways of adjusting to a new factor of the environment are possible. If a new predator should enter a biotic community, the various animals which might form its prey could gain added protection in a number of different ways. Some could acquire greater agility, others thicker skins or shells, others better ability for hiding under cover, and still others could develop noxious substances repellent to the new potential enemy.

In plants, a good example of this diversity of adaptation is provided by a weed of the mustard family (*Camelina sativa*) which has developed races capable of infesting the flax fields of Europe. The common form of this species is a branched, bushy annual with small white flowers, small seed pods, and rather small, rounded seeds (Figure 4-10). This form grows on roadsides and open fields. It is not adapted to infesting flax fields. Flax plants grow tall, straight, and very close together, so that a weed is successful in this habitat only if it also is tall, straight, and unbranched. Furthermore, the seeds of the roadside *Camelina* are easily kept out of flax fields, since their shape and size is so different from that of the large, flat seeds of flax that the two kinds of seeds are easily separated when the flax seeds are purified by winnowing. In regions of eastern and northern Europe where flax is commonly grown, the weedy *Camelina* has evolved races especially adapted to infesting fields of this crop. They are tall, unbranched, and have much larger capsules and seeds than do the ordinary races found along roadsides. The size and specific gravity of these weed seeds has been adjusted by natural selection to resemble those of flax, so that they are blown to equal distances by the winnowing machine and remain in the sample of flax seeds which the farmer sows.

The increase in seed size, therefore, is one of the most important adjustments by which *Camelina* becomes adapted to the flax field environment. The genes responsible for this adjustment, however, bring about simultaneously a loss in reproductive capacity. Genetic studies have shown that genes for seed size in *Camelina* automatically reduce the number of seeds per capsule. The number of seeds per plant is still further reduced by the necessary reduction of branching. Consequently, any genes which will make up for this loss in reproductive capacity will have a high selective value in the adaptive complex of flax weed *Camelina*. Such genes have become established. They act by increasing the size of the capsule without altering seed size. These genes have produced the most conspicuous differences which exist between different races of flax-infesting *Camelina*. The size of the capsules can be increased equally well by increasing their height, their width, or their thickness. Two different solutions to this problem of adaptation, both of which are common in Sweden, are shown in Figure 4-10 C and D. They have most probably originated separately, in different regions, from a similar, widespread, ancestral stock.

A final factor which may influence the differentiation of some populations is chance. In any population, characters which have relatively low adaptive values may oscillate in frequency. In one generation, there may be an exceptionally high proportion of matings between individuals homozygous for one of a pair of alleles. This temporarily increases the frequency of that allele in their progeny. This deviation will usually be compensated in a later generation by a chance increase in frequency of matings between individuals homozygous for the opposite allele. In large populations, these chance fluctuations in gene frequency cannot have any effect on evolution. A deviation in one direction on the part of a few individuals is certain to be balanced by a chance deviation in the opposite direction elsewhere in the population. If, however, a population is very small, one might expect that such fluctuations could occasionally lead to the complete loss of one allele from a population, and the consequent "fixation" of homozygosity for the opposite allele. In artificially constructed populations of *Drosophila*, which have contained as few as ten to a hundred flies, such fixation has been demonstrated.

Since this principle was first put forward by Dr. Sewall Wright more than thirty years ago, many efforts have been made to explain differences between natural populations on this basis. At present, however, no example of differentiation between populations is known which can be ascribed solely or principally to this factor. One reason is that the effects of selection, both direct and indirect, are very strong, particularly in populations which are becoming smaller because of a worsening environment. Consequently, no gene remains unaffected by selection for a sufficient number of generations so that its frequency can be permanently altered by chance alone.

On the other hand, chance factors may be associated with natural selection in small populations, and be responsible for their differentiation with respect to some characteristics which have little adaptive value. A model for this type of action is familiar to many people who have lived in country districts with relatively stable populations. This is the distribution of surnames in small villages. For instance, in a part of New England where the writer was brought up, a different and somewhat unusual group of surnames was characteristic of each village. In one village, most of the old settlers bore the name Jordan or Clement, in another Bracy and Hancock were the most common, while in a third Fernalds were predominant. This condition can probably be traced back to a few early settlers in each village who had a large number of male children and grandchildren. The names themselves had nothing to do with this greater fecundity, but became established merely by association with it. Because it is usually the outcome of the particular characteristics of a few founders of a new population, this type of association between non-adaptive and adaptive characteristics is known as the FOUNDER PRINCIPLE.

This interaction between selection and chance is probably responsible for many differences between populations of islands and mountain peaks in respect to characters which by themselves have little adaptive value, such as small variations of leaf shape in plants, and patterns of markings on the wings of insects. Such differences can be intensified by the degree of inbreeding which inevitably takes place in such populations. If such island or mountain populations later increase in size, their differences can become incorporated into differences between widespread species. Moreover, chance differences such as these may determine different initial ways of adapting to similar environments, which will become accentuated as the environment changes further. Chance factors by themselves probably have little effect on evolutionary processes, but in combination with reduction in population size and the accompanying natural selection, they may play an important role in guiding the early stages of adaptive differentiation.

The above discussion has focused attention on a question which was asked by some of the earliest critics of Darwin's theory of natural selection: How can natural selection account for the establishment of the first rudiments of a new adaptive mechanism, of which the initial advantage is relatively slight? One answer to this question was given by Darwin. He reasoned that since any increase in adaptive efficiency, however slight, will have some selective advantage, only a large enough number of genetically variant individuals and a long enough span of time are required to bring any kind of adaptive modification into being. A second answer is provided by the studies of the evolution of mimicry, previously described. They tell us that the evolution of a new adaptive gene complex can be triggered when a mutation or random fluctuation shifts the phenotype for a considerable distance in the direction of this adaptation and takes place in the presence of an environmental change favorable to it. Third, the chance association in a small founder population between a character having little adaptive value and another which is highly adaptive, accompanied by the effects of inbreeding, can set the stage for a new direction of adaptation.

A fourth way in which a new adaptive gene complex can begin its evolution is by the process known as GENETIC ASSIMILATION. This process, which takes advantage of the phenotypic plasticity possessed by most organisms, has been illustrated by the experiments of Professor C. H. Waddington. In one of these experiments, he found that some flies in a laboratory population

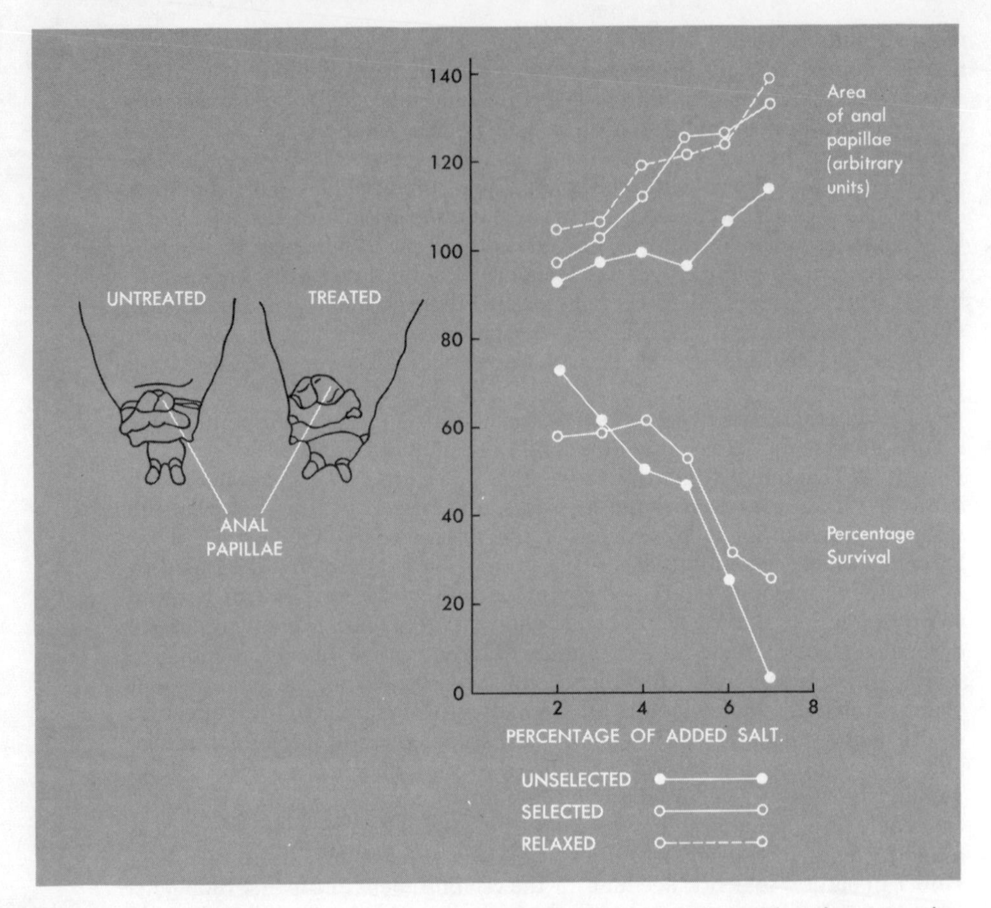

Fig. 4-11. *Left,* Drawings of the anal plates of larvae of *Drosophila melanogaster* when raised in normal medium (*left*) and in a medium with a high salt content (*right*). *Right,* graphs showing the response to increasing salt concentrations of normal lavae and of those selected for 21 generations in a medium with a high concentration of salt. Note that the selected lines (open circles) are able to modify their anal papillae phenotypically to a greater degree than can the unselected lines and also show a greater tolerance to high concentrations of salt. From Waddington, in *The Evolution of Life,* S. Tax Editor (Chicago University Press).

of *Drosophila* responded to heat shocks by producing wings lacking one of their cross veins. By selecting the flies with the strongest response, he was able to increase in later generations the percentage of flies which produced crossveinless wings even when not exposed to temperature shocks.

Although this experiment involved artificial selection for a non-adaptive character, Waddington reasoned that if an extreme environment could evoke an adaptive response of the phenotype, this could initiate selection for genes which under normal conditions would never express the adaptive character, and so could not normally be selected. Consequently, he exposed *Drosophila* larvae to high concentrations of salt in their diet, and found that they responded by an increase in the size of their anal plates (Figure 4-11), which aided in excretion of the harmful material. Those larvae which had the greatest phenotypic response produced the healthiest, most fecund adults, and thus brought about natural selection for intensification of the response. The population of flies produced after several generations of selection in this medium had both a higher salt tolerance and larger anal plates. They retained this phenotype when raised in media containing either high or low concentrations of salt.

Although this last experiment appears superficially like a verification of Lamarck's theory of the inheritance of acquired modifications, Waddington's two experiments taken together exclude this possibility. In the first experiment, the expression of the crossveinlessness phenotype under normal conditions was clearly produced by artificial selection of preexisting genetic differences. In the second experiment, genetic differences in the ability to respond to high salt concentrations also existed in the original population, and there is every reason to believe that if these differences had not been present, the experiment would not have succeeded. Only two assumptions are needed to explain the two experiments. The first is that the characters concerned are determined by many different pairs of genes. Second, we must assume that genes, which by themselves or in small numbers can produce a character only in an environment especially favorable for its phenotypic expression, can produce the same character under any environment if they are aided by several other genes having a similar action. This assumption is quite reasonable on the basis of known relationships among genes, enzymes, and metabolic processes. We might postulate that a particular phenotypic character will appear only when a required substance is present at a concentration higher than a necessary threshold (Figure 4-12). The production of this high concentration depends upon the simultaneous action of several genes, each of which by itself can produce only a sub-threshold amount of

Fig. 4-12. Chart showing a way in which genetic assimilation can be explained in terms of the additive action of genes which determine specific and similar enzyme reactions, together with the effect of an extreme environment in lowering the threshold necessary for the expression of a particular phenotypic characteristic.

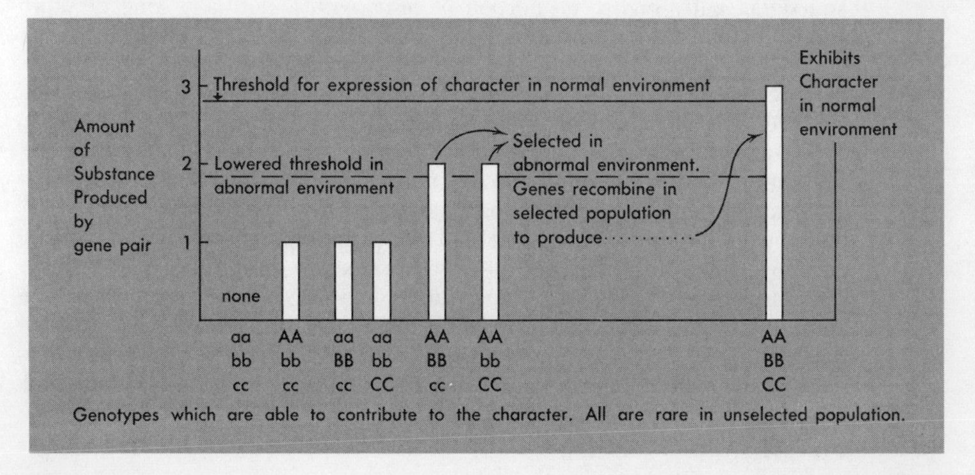

the substance. The environmental stress, acting in the same manner as temperature shock or high salt concentration, lowers the threshold so that even small amounts of the necessary substance can produce the phenotypic effect. Selection, either artificial or natural, can then pick out and accumulate the otherwise concealed genes, raising their frequency in the population to a level high enough for their concerted action to surmount the high threshold existing in the normal environment.

TYPES OF NATURAL SELECTION

From the discussion already presented in this chapter, the reader should realize that the effect of natural selection upon the course of evolution depends upon the way in which the environment is changing relative to the adaptive needs of the population. Based upon different organism-environment relationships, three different kinds of selection have been recognized, STABILIZING SELECTION, DIRECTIONAL SELECTION, and DISRUPTIVE SELECTION.

Stabilizing selection is the action of natural selection in keeping a population genetically constant. It occurs whenever the organism-environment relationship remains constant for long periods of time. The environment does not need to remain constant in all of its features, and in fact rarely if ever does so. Any particular population of organisms, however, takes advantage only of certain parts of the environment and does so in a particular way. Consequently, the organism-environment relationship will be changed only if those features of the environment are altered which affect the population concerned. The significance of this fact in relation to the different rates of evolution which have been recorded in the fossil record of past organisms will be brought out in a later chapter.

Stabilizing selection maintains genetic constancy by favoring average or normal individuals of a population, and eliminating extreme variants. A well-known illustration of its direct action was provided by H. C. Bumpus, who in 1899 measured the size and proportions of those individuals of a flock of sparrows which had been killed in a severe storm. He found an unexpectedly and significantly higher proportion of birds with abnormally long or short wings, relative to the average for the population.

Many studies of variation in natural populations some of which have been reviewed in this chapter, have shown that stabilizing selection does not act by establishing and maintaining an optimal "type," consisting of individuals which are all alike and homozygous for those genes most concerned with adaptation. Rather, the optimum or "norm" of adaptation is represented by a cluster of genotypes which possess different gene combinations, many of them in the heterozygous condition. Their phenotypic resemblance is due in part to genetic similarity, and in part to selection during past generations. This has constantly weeded out individuals deviating phenotypically from the norm, regardless of the means by which these deviants have arisen. A population existing in a constant environment (relative to its needs) is therefore in a state of dynamic equilibrium between the stabilizing force of selection and the disruptive forces of genetic segregation and mutation.

Evidence that this condition exists is provided partly by the fact that artificial selection can cause a population to diverge rather quickly from its adaptive norm, provided that a mutation or an extreme segregant individual gives the selector a start. Such rapid selection for deviants has been carried out on the crooked toe character in poultry, on extra numbers of bristles on the scutellum of *Drosophila* flies, and on deviating numbers of facial whiskers in the mouse. In each case, however, the selected strain was much less viable than the population from which it was obtained. Artificial selection had shifted the population away from its adaptive norm.

In another experiment, the stabilizing action of natural selection on such a deviant population was demonstrated experimentally. In the red checker moth (*Panaxia dominula*) artificial selection was carried out for an abnormal phenotype, pale wings, and after several generations an unusually pale-winged strain was obtained which bred true. This strain was then released in a locality in Britain where no moths of this species existed naturally. Because the species is very sedentary in its habits, the investigators could be reasonably sure that the descendants of the released moths were not able to cross with wild individuals of the species. Nevertheless, after five years the descendants of these moths had formed a population with normal wing coloration. Natural selection had favored genetic combinations and mutations which shifted the wing coloration back to the adaptive norm for the species.

Directional selection produces a regular change of the population in one direction in respect to certain adaptive characteristics. It occurs when the environment is changing progressively in a particular direction. Deviants from the norm in one direction tend to survive more often and to produce more offspring than do deviants in the opposite direction. This is what happened during the selection of the colon bacterium for resistance to increasing doses of puromycin (described at the beginning of this chapter). Deviants (in this case mutants) with greater resistance to puromycin had a better chance of surviving and producing offspring, while susceptible cells, even those which reproduced more rapidly in a puromycin-free medium, were destroyed in greater numbers in the environment containing increased concentrations of the drug.

Under natural conditions, directional selection might be expected under two different sets of conditions. It would occur in a population subjected to a progressive change in the environment. This could consist, for example, of an increasingly cold or dry climate, or of the attacks of a particular kind of predator which was itself evolving increasingly more efficient methods of predation. It could also consist of a progressive improvement in the environment, which would favor selection for greater reproductive potential. Evidence of such evolutionary trends in the fossil record will be presented in a later chapter. Directional selection would also occur in a population which was migrating into a new territory having progressively altered environmental conditions. For instance, at the end of the last ice age many species migrated northward into the regions laid bare by the retreat of the glaciers. They must have been subjected to increasing cold, and an increasingly extreme cycle of contrast between the short days and long dark nights of winter and the long days and short nights of summer. Only those populations capable of adaptive responses to these conditions would have been able to migrate a long distance northward. Genetic tests under uniform conditions of the various species of grasses inhabiting our Great Plains have shown that they consist of series of populations which vary genetically in a regular fashion as one goes northward or westward from the Mississippi Valley and the coastal plain of Texas. These variations adapt the plants to increasing rigors of cold and drought. A similar study of pine seedlings grown from seed collected in all parts of Sweden has shown increasing adaptations to cold in the strains derived from increasingly northern localities.

Disruptive selection acts to break up a previously homogeneous population into several different adaptive norms. It occurs when a population previously adapted to a homogeneous environment becomes subjected to divergent selection pressures in different parts of its area. An example of disruptive selection was found by the writer and his coworkers in the history of a population of sunflowers in the Sacramento Valley of California over a period of twelve years. When first discovered, this genetically variable population, produced from a hybrid between two species, was occupying an area about 300 feet long and 20 feet wide, in a ditch. Five years later, it had split into two sub-populations, separated by a grassy area about 500 feet long in which few sunflowers could grow. One of these, which occupied a relatively dry site, had diverged in the direction of one of the parents of the hybrid. The other, which occupied the bottom of a ditch that remained wet until late spring, retained the general variability of the original hybrid swarm with slight divergence in the direction of the other parent. During the next seven years the size of the population fluctuated greatly in response to differences in rainfall from one season to the next, but the differences between the two sub-populations were maintained. These annual sunflowers produce one generation a year. They cannot be fertilized by their own pollen and are normally cross pollinated by bees. Bees could easily fly from one sub-population to the other, so exchange of genes between them undoubtedly occurred. However, the divergent forces of selection in the different habitats occupied by the two sub-populations were strong enough to keep them genetically distinct.

Differential response to the same environmental change can take place, as Professor Sewall Wright has pointed out, when a population is subdivided into a number of small sub-populations, each one having a slightly different gene pool. This is the usual condition of species which occupy specialized and restricted habitats, such as swamps, pond margins, steep cliffs or talus slopes, sand dunes near the sea, or cultivated fields of a particular type. An example of disruptive selection of this type, given earlier in this chapter, is the evolution of flax-inhabiting races of *Camelina* having capsules of different shapes.

The discussion of this chapter is intended to show that the principal guiding force of evolution is the organism-environment interaction, which produces genetic changes in populations through the operation of natural selection. Because of the enormous complexity of both organisms and their environment, this interaction can alter populations in a multitude of different ways. The indirect action of natural selection is at least as common and as important to evolution as are its direct, obvious effects.

Chapter Summary

The action of natural selection under controlled conditions can be followed most easily in cultures of bacteria exposed to gradually increasing concentrations of antibiotics. They can become resistant to high concentrations of these drugs through the occurrence and establishment of a succession of mutations toward resistance. The establishment of melanic mutations in natural populations of moths in association with the industrialization of certain parts of northern Europe, a phenomenon known as industrial melanism, is the best example of historically recorded changes in populations in response to alteration of known factors in the environment. These and many other recorded changes show that natural selection does not necessarily act through a direct struggle between individuals. Usually, the successful individuals are better able to take advantage of certain factors of their environment, and consequently produce a larger number of offspring. In this way natural selection can be effective even if no adult individuals are killed or die and if propagules and young offspring are destroyed at random.

Among the most elaborate adaptations which have been established by natural selection are those constituting mimicry in butterflies and other animals. Their effectiveness has been demonstrated by a series of carefully controlled experiments. In addition, the relative frequencies of mimicking, nonmimicking, and imperfectly mimicking phenotypes have been determined in natural populations of the African butterfly, *Papilio dardanus,* while their genetic basis has been revealed through hybridization. This research has shown that mimicry can evolve through the initial appearance of a mutant form which is an imperfect mimic, followed by the occurrence and establishment of modifying genes which perfect the mimicry.

Differences which appear to have no adaptive value can become established by natural selection in the following ways: (1) An adaptive difference exists but is not apparent; (2) The difference is controlled by a gene or genes with pleiotropic effects, some of which are visible and non-adaptive, while others are invisible and adaptive; (3) A non-adaptive difference may be genetically linked with an adaptive one; (4) the difference may represent two alternative and equivalent ways of adapting to the same environmental change. In small populations, non-adaptive differences can become established through chance association with adaptive characters. If the populations later increase in size, the difference can be maintained. This is known as the FOUNDER PRINCIPLE.

One way in which selection can be brought to bear on characters which in their initial stages are weakly expressed is through the strengthening of this expression by phenotypic modification in extreme environments. This is followed by the accumulation of enough favorable genes in the population to enable the character to be expressed in any environment. This process is known as GENETIC ASSIMILATION.

Three different modes of action of natural selection are STABILIZATING SELECTION, which occurs when a population is exposed to a basically constant environment; DIRECTIONAL SELECTION, which is produced by a continuous change of a population in a particular direction; and DISRUPTIVE SELECTION, which results from the exposure of different parts of the same population to divergent selective pressures.

Questions for Thought and Discussion

1. Of the three primary processes governing the rate and direction of evolution (mutation, recombination, selection), would you consider any one to be more important than the others? If so, why? If not, why not?

2. What are the advantages and disadvantages of using microorganisms as models to illustrate the general principles of natural selection?

3. Explain why the existence of adaptive superiority in a genotype or gene complex is easier to demonstrate experimentally than are the reasons why the superiority exists.

4. Explain how natural selection can lead, over several generations, to the complete elimination of one type and the establishment of another even though in each generation no individual dies between birth, hatching, or seed germination and reproductive maturity.

5. Using examples, explain how chance events may affect the outcome of selection.

CHAPTER 5

Reproductive isolation and the origin of species

In the last chapter the fact was brought out that populations evolve new characteristics largely as a result of changes in the organism-environment relationship, acting through the medium of natural selection. Since environments are very diverse and complex in both space and time, organism-environment relationships are always different for related populations living in different parts of the earth, and are perpetually changing as the earth's environments are altered in various ways. Changes in climate, in the distribution of land and sea, elevation and degradation of mountain systems, and particularly the increase, decrease, and extinction of different kinds of animals and plants all interact to produce these changes in organism-environment relationships. The remaining chapters of this book will be devoted to showing how these changes affect the nature, course, and rate of evolution.

The general sequence of events, which in the next two chapters will be described and analyzed with examples, can be outlined as follows. We can start with a single relatively homogeneous population, living in a particular area and possessing a reasonably large pool of genes, from which combinations adapting some of its individuals to new conditions can be sorted out by natural selection and established in new habitats. If this condition exists and if new habitats are available, the evolutionary line derived from the initial population will undergo ADAPTIVE RADIATION. Individuals entering new regions will found subpopulations which will differ in various adaptive characteristics from the ancestral population and from each other. In this way a POLYTYPIC SPECIES will evolve consisting of many races adapted to different conditions. The most easily recognized of these races, or groups of races, are usually called SUBSPECIES. From time to time, but by no means always, races can diverge from their ancestral populations in ways which decrease or prevent their ability to exchange genes with other races. This may come about either by reducing the frequency with which interracial hybrids are produced or by lowering the viability or fertility of the hybrids or their progeny.

When any of these barriers to gene exchange, known as ISOLATING MECHANISMS, have become strongly developed, the populations are said to be REPRODUCTIVELY ISOLATED from each other. They then rank as distinct SPECIES. If, thereafter, new environmental changes enable reproductively isolated populations to migrate into the same region, they can exist side by side, each population maintaining its own distinctive adaptive gene complex. Each of these newly evolved species can undergo its own distinctive cycle of adaptive radiation and speciation. In this way an adaptive radiation develops which is of a higher order than that which gives rise to a polytypic species. This results in the evolution of clusters of related species which form a GENUS.

Sometimes two differently adapted populations or subspecies of a polytypic species can come together before reproductive isolation between them has been fully developed. In that case they can exchange genes to a limited degree, and progeny resulting from such HYBRIDIZATION may contain individuals adapted to environments different from those occupied by either of the

Fig. 5-1. Diagram showing the sequence of events which leads to the production of different races, subspecies, and species, starting with a homogeneous, similar group of populations.

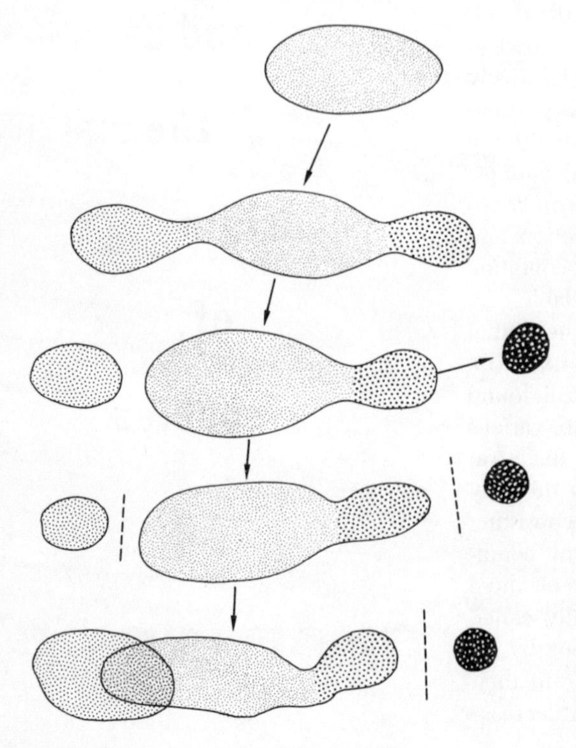

First stage.
A single population in a homogeneous environment.

Second stage.
Differentiation of environment, and migration to new environments produces racial differentiation of races and subspecies (indicated by different kinds of shading).

Third stage.
Further differentiation and migration produces geographic isolation of some races and subspecies.

Fourth stage.
Some of these isolated subspecies differentiate with respect to genic and chromosomal changes which control reproductive isolating mechanisms.

Fifth stage.
Changes in the environment permit geographically isolated populations to exist together again in the same region. They now remain distinct because of the reproductive isolating barriers which separate them, and can be recognized as good species.

parental populations. Although such hybrids are most often unsuccessful because of their weakness or sterility, some of them may give rise to new populations, well adapted to a new habitat which may be available. Under such exceptional circumstances, hybridization can greatly speed up the processes of evolution. In plants, hybrids may give rise to new species by doubling their chromosome number, a process known as POLYPLOIDY. In this way, a sterile hybrid can give rise immediately to fertile offspring, which will remain constant for the intermediate, hybrid condition.

This sequence of events is illustrated in Figure 5-1. The five stages shown in this diagram and explained in the legend, are somewhat oversimplified, but nevertheless give a general idea of the commonest way in which species can originate.

RACIAL DIFFERENCES WITHIN SPECIES

Nearly every species is differentiated into races, and in many of them there exist races distinctive enough to be called subspecies. For example, a person familiar with the birds of the eastern United States can recognize the song sparrow (*Melospiza melodia*) by its size, color pattern, and song (Figure 5-2). If he travels to the west coast of our continent, he will hear a song which reminds him of that sung by the eastern song sparrow but is not quite the same. When he sees the bird which is singing this song he will recognize still further differences. Furthermore, the song sparrows of the Pacific Coast are not all alike. Those of central California are recognizably different from the song sparrows of Oregon and Washington. Within the San Francisco Bay region, populations of song sparrows inhabiting hill slopes and marshes separated from each other by only a few miles are recognizably different.

We might ask at once, "Doesn't this mean that there are actually several different species which are all given the common name of song sparrow?" The biologist at once answers, "No, these different kinds of song sparrows are no more different species than are the various races of mankind." The reason for this answer is that wherever areas of geographic distribution of two different races or subspecies of song sparrow come close to each other, intermediate populations are found. One can never find two or more different races of this bird breeding in the same locality and remaining distinct from each other.

Species of plants also consist of many different populations. If we look at the pine trees which cover the slopes of a high mountain, we notice that those near the summit are much smaller than those near the base, and they are likely to be gnarled and spreading. The same is true of trees and shrubs which grow near the seashore and are exposed to constant winds. We might suspect that these differences result directly from the effects of the severe environment. This is true to a certain extent. Nevertheless, when seeds are taken from trees growing at various altitudes in the mountains, and their offspring are raised in a uniform nursery, those descended from trees growing at the lower altitudes grow faster and become larger than do the progeny of trees which were growing at higher altitudes. Evidently, therefore, natural selection has sorted out genotypes adapted to the conditions of their native environment. Such adaptive differentiation is not confined to size and rate of growth. For instance, many smaller plants or herbs survive the harsh winters of northern climates or high altitudes by dying back to their roots and becoming dormant during the winter months. If such plants are transported to a mild climate where little or no frost occurs, they will, nevertheless, become at least partly dormant during the winter. Related races of the same species which are native to the mild climate will grow actively during this same period.

In a number of instances, hybrids have been made between such differently adapted races or subspecies of plants. The hybrids are usually vigorous and intermediate in most respects between the parental races. When seeds are grown from such hybrids, they produce a large and extremely variable segregating progeny, no two individuals of which are genetically

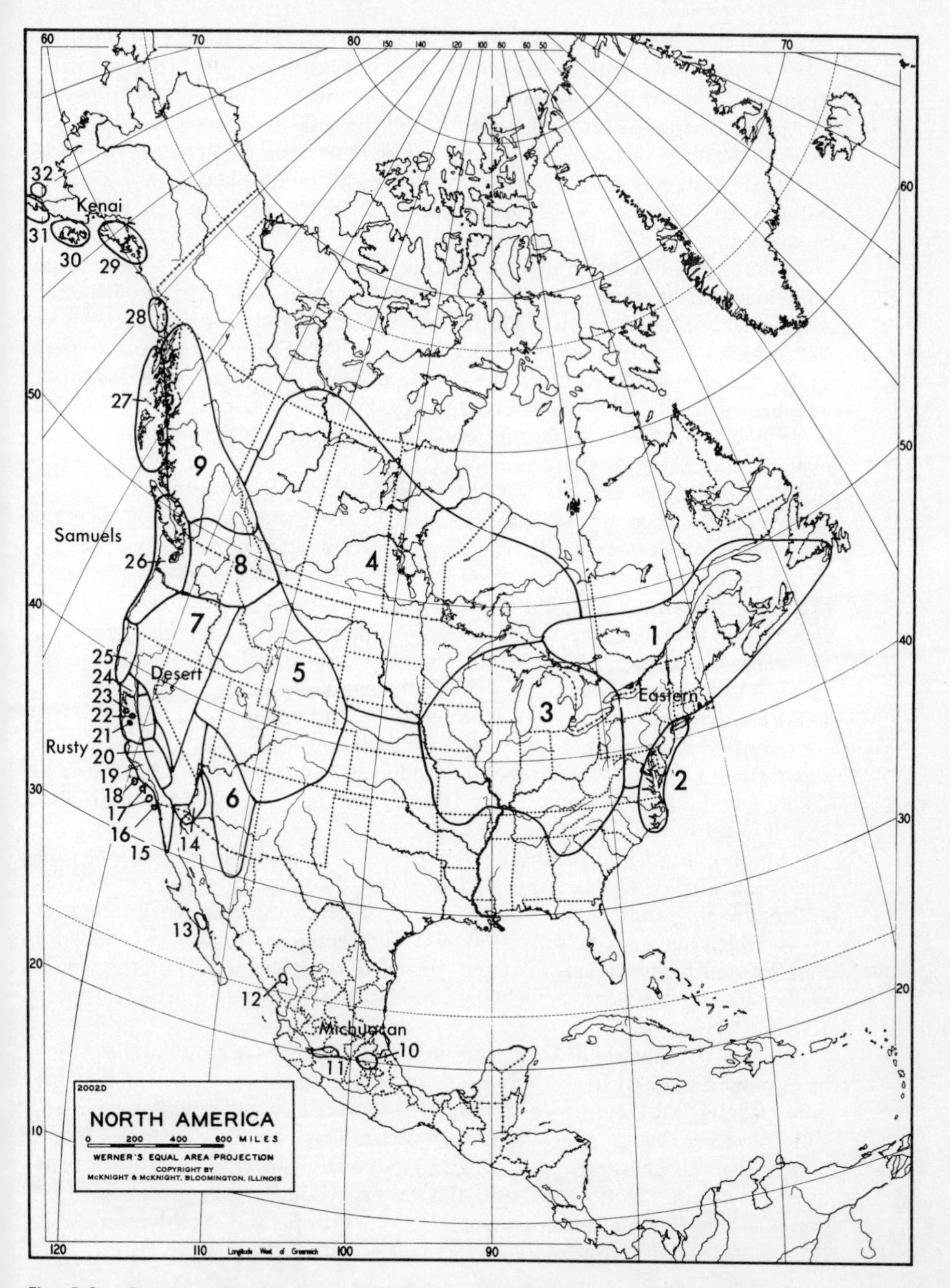

Fig. 5-2. Representative subspecies of the song sparrow (*Melospiza melodia*) together with a map showing the distribution of the numerous subspecies in North America.

alike. Furthermore, when such progeny are grown in a variety of environments, individuals can be recognized to have adaptive properties different from both parents, as well as from the F_1 hybrid. The segregating progeny of a hybrid between two differently adapted races of a plant species form a hybrid swarm, containing a greatly increased pool of genes controlling adaptive characteristics. If such hybridization takes place extensively in nature, the adaptive complexes of the parental populations can rarely be reconstituted in their original form.

Sometimes the differentiation of races within a species is not associated with marked differences in the climate or habitat which they occupy. For instance, the island of Celebes in Indonesia is inhabited by six different races of a species of carpenter bee (*Xylocopa nobilis*). These races inhabit very similar rain forests. The differences between them, though probably adaptive, are not related to any easily recognizable differences in their habitat. Quite possibly, they represent different modes of adaptation to the same habitat, as explained at the end of the previous chapter.

Although most populations of a species inhabiting different regions are recognizably different from each other, this is not necessarily so. Several species of plants inhabiting two or more widely separated areas have populations so similar in the two areas that the difference between areas is probably less than the variation within a population or between neighboring populations. An example is the skunk cabbage (*Symplocarpus foetidus*), common in swamps throughout the eastern United States and found also in similar places in China. In these examples, geological and paleontological evidence indicates that exchange of genes between the separated populations has not been possible for millions of years. Consequently we must conclude that separation of populations merely by distance is not enough to cause them to diverge and form different races. They must be subjected to differential selective pressures.

Some examples of mixing between races tell us, furthermore, that race formation is a reversible process. Races can disappear through hybridizing and combining their genes with those of other races. This is true even of races which have been separated and differentiated from each other for thousands or even millions of years. Among birds, this is shown by the woodpecker-like birds known as flickers (*Colaptes auratus*) in the northern United States.

Among woody plants there are numerous examples of populations which have been separated from each other for millions of years and have become differentiated into entities generally recognized as species, but which are still able to exchange genes with each other. This condition has been found in pines, oaks, chestnuts, plane trees, and many others. For instance, the hybrid between the plane tree of the eastern United States (*Platanus occidentalis*) and its counterpart in southeastern Europe and southwestern Asia (*P. orientalis*) is often grown in parks and is fully fertile.

SPECIES AS DISTINCT SYMPATRIC POPULATIONS

The facts presented above tell us that related populations can be separated from each other for long periods of time, extending even to millions of years, and can be subjected to such different selection pressures that they come to differ markedly from each other while still separated. Nevertheless they can retain the ability to exchange genes freely and reunite into a single interbreeding population if they are brought together again. There exist, however, many groups of related populations which live together in the same area, but which remain distinct from each other without exchanging genes. These are the entities which we usually recognize as species. Anyone familiar with the animal life of the lakes and streams of the northeastern United States can recognize the leopard frog (*Rana pipiens*), the wood frog (*Rana sylvatica*), or the bullfrog (*Rana catesbiana*). Although these species occur commonly in the same area, they never intergrade. Some acquaintance with birds will enable one to distinguish between the song sparrow (*Melospiza melodia*) and the Lincoln's sparrow (*Melospiza lincolni*) (Figure 5-3). Populations of these species live together over considerable areas of our country and regularly remain distinct from each other. They are apparently incapable of exchanging genes under natural conditions. The same is true of such easily recognized species of plants as the sugar maple (*Acer saccharum*) and red maple (*A. rubrum*) (Figure 5-4), long leaved pine (*Pinus palustris*) and scrub pine (*P. virginiana*), as well as ponderosa pine (*P. ponderosa*) and digger pine (*P. sabiniana*) in California.

At this point, the reader should become fully aware of what is meant by the phrase "incapable of exchanging genes under natural conditions." It does *not* mean that species can become distinct from each other only by evolving mechanisms which prevent them from hybridizing under any conditions, or which render the hybrids completely sterile. There are many examples of perfectly distinct and valid pairs of species which nevertheless can form fertile hybrids under artificial conditions. For instance, the common mallard duck (*Anas platyrynchos*) and the pintail duck (*Anas acuta*) will hybridize readily and produce fertile offspring when kept in captivity, but due to their

Fig. 5-3. Two related but sympatric and distinct species of North American birds, the Lincoln sparrow (left) and the song sparrow (right). From R. T. Peterson, *Field Guide to Western Birds* (Houghton Mifflin Company).

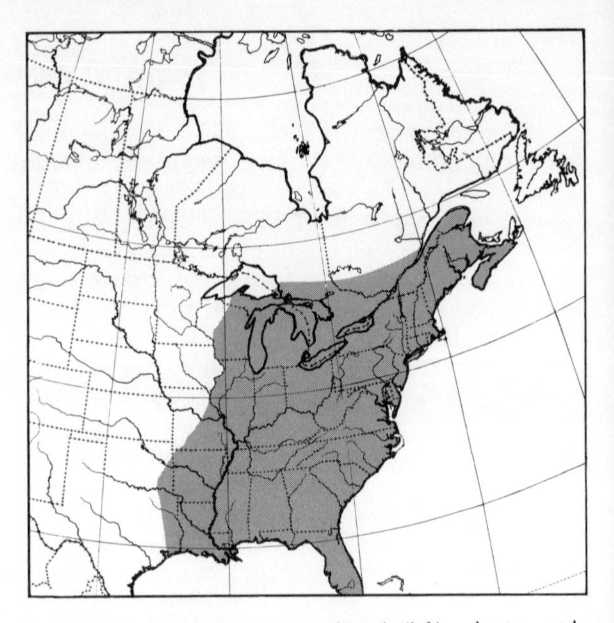

Fig. 5-4. Two related and sympatric species of trees, the red maple (left) and sugar maple (right). From Preston, *North American Trees* (Iowa State University Press).

very different courtship and nesting habits, they do so very rarely in nature. Hence they can maintain themselves as distinct sympatric species, even though they nest side by side in millions of ponds or creeks.

One of the basic questions asked by the evolutionist is: What factors in the evolutionary history of these related populations have made them diverge to such an extent that they are no longer capable of exchanging genes? This question is so important that the remainder of the present chapter will be devoted chiefly to discussing possible answers to it. Its importance lies in the following fact. As long as populations retain their ability to exchange genes, this mixing will cause them to merge into a single population whenever they are brought together after separation, and will prevent them from diverging further. Consequently, they cannot become the progenitors of new and distinctive lines of evolution. If, on the other hand, populations have evolved distinctive characteristics of the sort which prevent them from exchanging genes with each other, they will remain distinct when brought together in the same region. Furthermore, natural selection will tend to reinforce this distinctness for the following reason. Each population can be assumed to have a gene pool of variability, which will include some variants more similar to and others more different from the individuals of the other species. Those more similar to the other species will be more likely to come into direct competition with it for food and for places to hide from predators. They will, therefore, be at a reproductive disadvantage compared to those individuals of the population which are more different and are beginning to exploit their environment in a different way. Natural selection will, therefore, promote

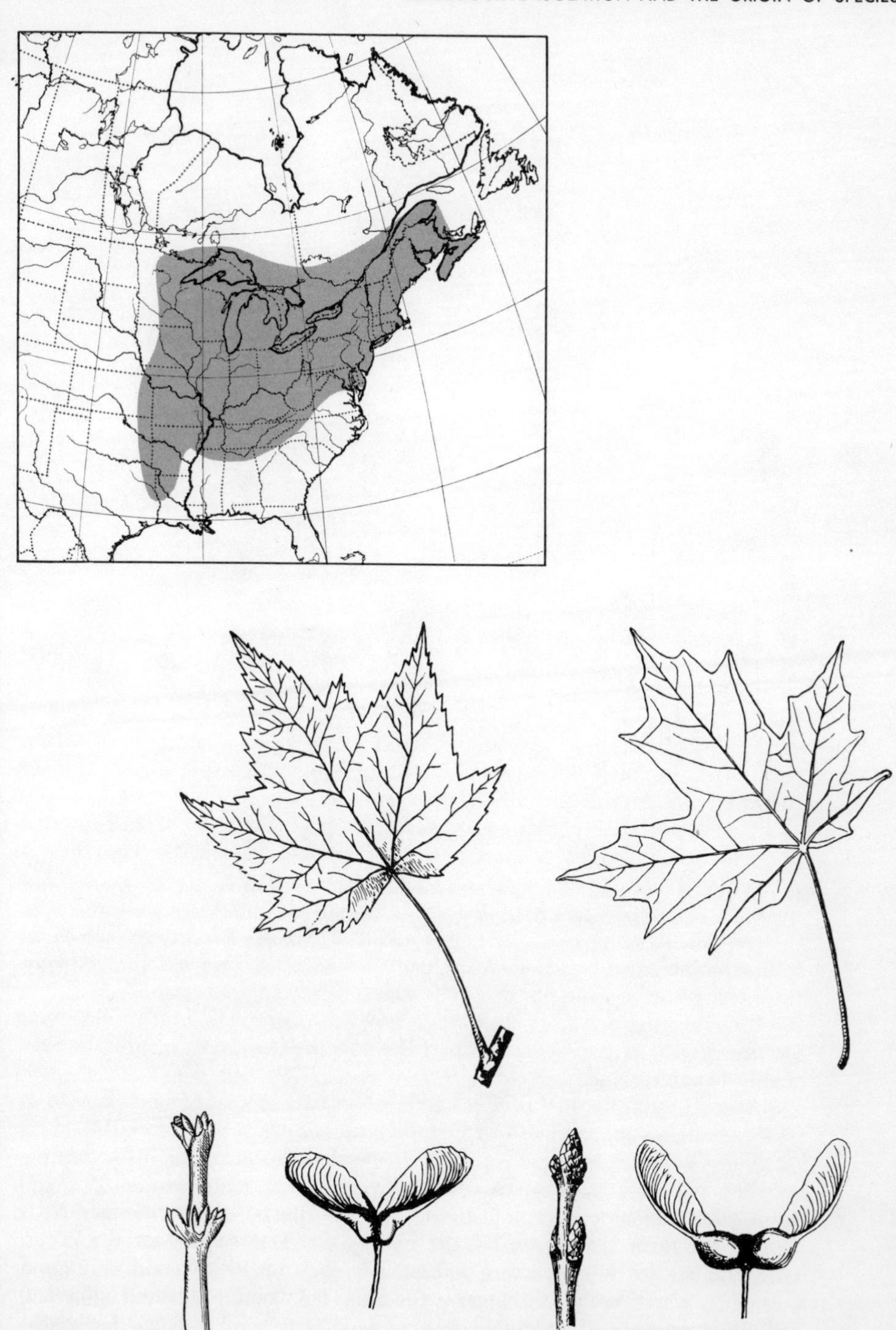
Red Maple
Sugar Maple

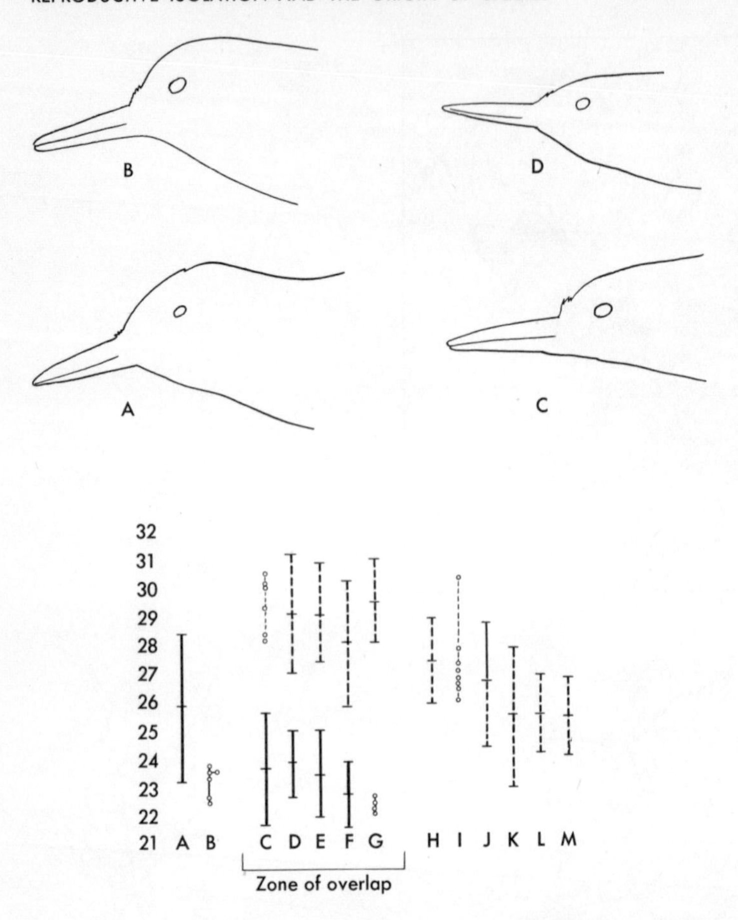

Fig. 5-5. Diagrams showing character displacement in respect to beak size in two species of nuthatches (A-C and B-D) which inhabit southeastern Europe and southwestern Asia. In the region where the two species occur sympatrically (C, D) they differ from each other to a greater extent than in regions where only one species is found (A, B). From Vaurie, in Proceedings Tenth Int. Ornith. Congr. Uppsala, 1950, p. 163.

the divergence of two species which live together in the same area but are unable to interbreed.

A good example of this effect is found in two species of birds known as rock nuthatches in southeastern Europe and western Asia (Figure 5-5).

The discussion in this chapter has revealed the need for distinguishing between populations or species which live together in the same area, and those which inhabit different areas. For this purpose, evolutionists have devised the term SYMPATRIC for the former and ALLOPATRIC for the latter. Thus, the eastern song sparrow and Lincoln sparrow, as well as the red and sugar maples, are sympatric species. The northern wood frog (*Rana sylvatica*) and the western yellow-legged frog (*Rana boylei*), as well as the southeastern yellow pine (*Pinus echinata*) and the western ponderosa pine (*P. ponderosa*) are allopatric species (Figure 5-6). You must recognize that the difference between sympatry and allopatry is not absolute, but one of degree. There are many examples of related species which are allopatric in some parts of their distributional ranges, but sympatric (i.e. overlapping) in others. A typical example is that of the two large coned pines, digger (*P. sabiniana*) and Coulter (*P. coulteri*) in California (Figure 5-7).

Because of its importance in promoting further divergent evolution and because the diversity of the flora and fauna inhabiting a region could not have evolved without it, the ability of two populations to exist sympatrically without losing their identity through hybridization and gene exchange is regarded by evolutionary biologists as being of prime importance in recognizing them as distinct species. When the related populations of an area are considered, those which "pass the test of sympatry," that is, those which retain their identity even though they live close enough together so that cross fertilization between them is possible, are regarded as separate species. On the other

Fig. 5-6. Two allopatric and related species of frogs, *Rana sylvatica* (northward and eastward) and *R. boylei* (California and Oregon). From R. C. Stebbins, *Amphibians and Reptiles of Western North America* (McGraw-Hill Book Company).

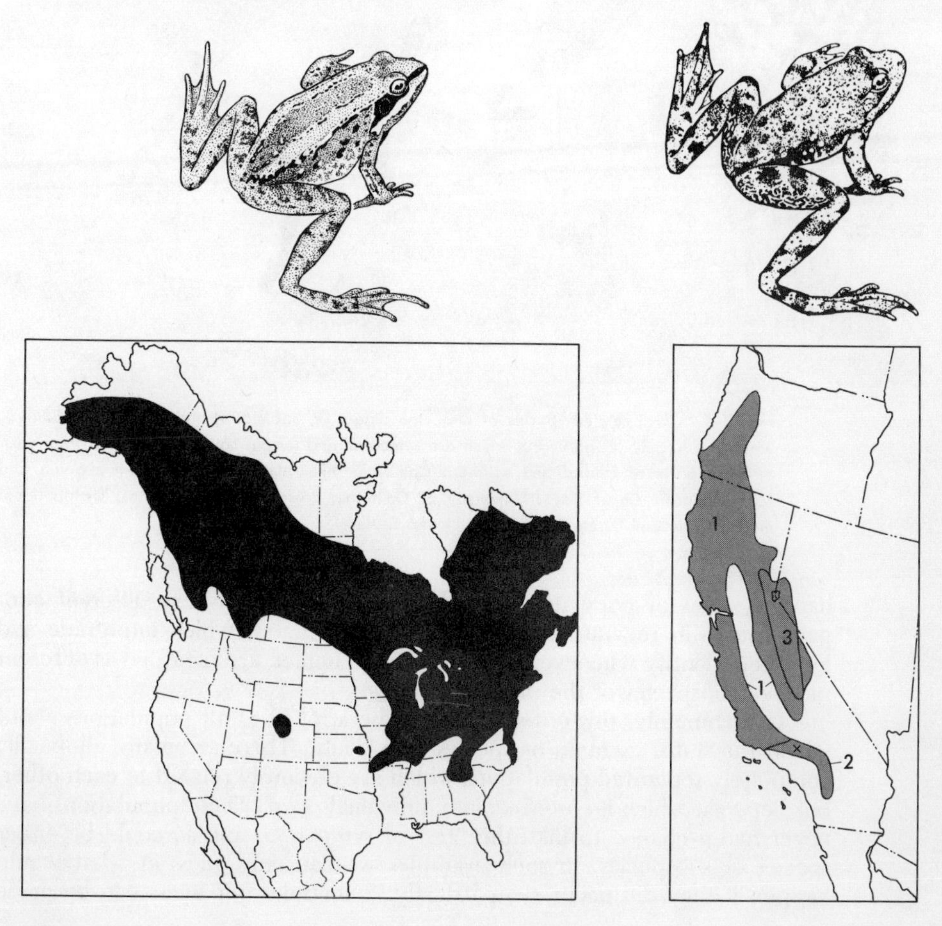

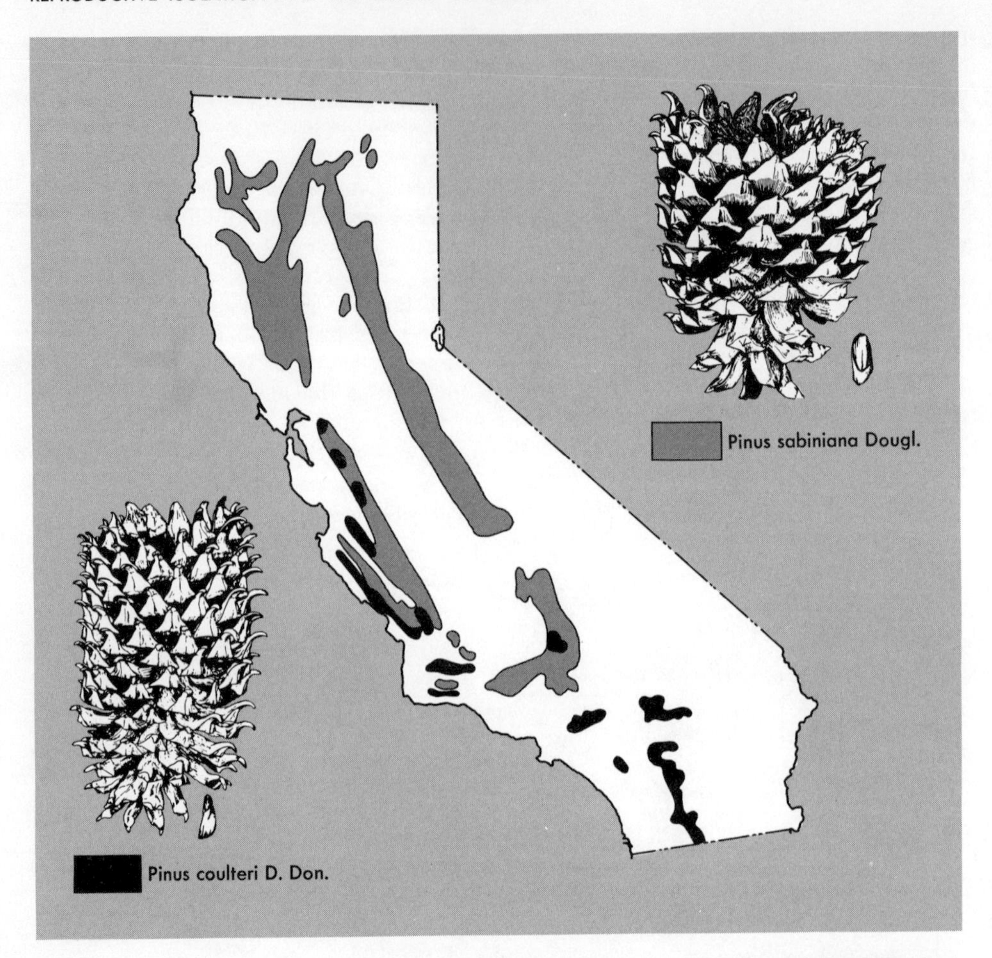

Fig. 5-7. Two related species of pine, the digger (*P. sabiniana*) and Coulter (*P. coulteri*), which are largely allopatric but which are sympatric and remain completely distinct from each other in parts of central and southern California. Illustrations of cones from McMinn and Mainio, Pacific Coast Trees, University of California Press; map from Munns, Distributional maps of principal Forest Trees of the United States.

hand, systems of populations which may have evolved very different allopatric races in regions remote from each other, but which intergrade and lose their identity whenever they come into contact, are regarded as different races or subspecies of the same species.

Unfortunately, this criterion cannot be applied to all populations about which the status as races or species is in doubt. There are many allopatric, completely separated populations which are obviously related to each other, but between which no intermediate individuals exist. These populations have never had a chance to take the "test of sympatry," and so could be either species or subspecies. In such examples we can only guess at what would happen if they did occur sympatrically. Sometimes our guess can be made more intelligent by bringing individuals of the two populations together and attempting artificial hybridization between them. There are other situations in which sympatric or partly sympatric systems of populations show varying degrees of distinctness or intergradation in the places where they occur together. If we assume that the origin of species is a gradual process (an assumption for which, as we shall see, there is much evidence), we would expect to find such intermediate degrees of distinctness between populations. They will be discussed more fully in the next chapter.

THE NATURE OF REPRODUCTIVE ISOLATION

The next question which we must ask is: What factors act to enable related, sympatric populations to keep their identity and remain distinct from each other? These factors are numerous and varied. Collectively, they are designated by the term ISOLATING MECHANISMS. The principal ones are listed in Table 5-1.

TABLE 5-1. Summary of the Most Important Isolating Mechanisms Which Separate Species of Organisms.

A. *Prezygotic mechanisms.* Prevent fertilization and zygote formation.

1. *Habitat.* The populations live in the same regions, but occupy different habitats.
2. *Seasonal or temporal.* The populations exist in the same regions, but are sexually mature at different times.
3. *Ethological* (only in animals). The populations are isolated by different and incompatible behavior before mating.
4. *Mechanical.* Cross pollination is prevented or restricted by differences in structure of reproductive structures (genitalia in animals, flowers in plants).

B. *Postzygotic mechanisms.* Fertilization takes place and hybrid zygotes are formed, but these are inviable, or give rise to weak or sterile hybrids.

1. *Hybrid inviability or weakness.*
2. *Developmental hybrid sterility.* Hybrids are sterile because gonads develop abnormally, or meiosis breaks down before it is completed.
3. *Segregational hybrid sterility.* Hybrids are sterile because of abnormal segregation to the gametes of whole chromosomes, chromosome segments, or combinations of genes.
4. *F_2 Breakdown.* F_1 hybrids are normal, vigorous, and fertile, but F_2 contains many weak or sterile individuals.

Isolating mechanisms can be classified into two groups, prezygotic and postzygotic. Prezygotic mechanisms are those which either prevent contact between the species when reproductively active, or which prevent or restrict the union of gametes after mating or cross pollination has occurred. Postzygotic mechanisms are those which prevent the growth of hybrid individuals after fertilization has occurred, or which reduce the fertility of the F_1 hybrids or the viability of their descendants.

PREZYGOTIC MECHANISMS

Habitat isolation is most common in plants because of their sedentary nature. It operates in two ways. The species may live in the same general area, but have such different habitat preferences that their populations are rarely close enough together to cross fertilize each other frequently. In addition, if hybrids are occasionally formed, they may be either unable to grow to maturity or leave few progeny under natural conditions, because no site to which they are adapted is available. Many examples can be found among the species of trees and shrubs found in the eastern United States. A good one is that of the scarlet oak (*Quercus coccinea*) and black oak (*Q. velutina*) (Figure 5-8). These two species, which are sympatric throughout most of the eastern United States, are easily told apart by the shapes of their leaves and acorns. Intermediate trees, although occasionally found, are usually much less common than are trees typical of the scarlet or the black oak, so that the two species populations retain their identity throughout nearly all of their vast area of distribution. This is principally because the scarlet oak inhabits swamps or poorly drained bottom lands having acid soils, while the black oak is found on drier, well-drained soils of the uplands. In some places, such as the north shore of Long Island, New York, intermediate habitats are relatively abundant, and the distinctness of the species breaks down.

In many instances, mating between sympatric populations is prevented or restricted because they reach reproductive maturity at different seasons of the year, a condition known as seasonal isolation. Since, however, seasonal differences can very greatly from year to year, this type of isolation is relatively ineffective by itself and is most important when combined with other types of isolation.

In animals the most important group of prezygotic isolating mechanisms are those associated with different behavior patterns in courtship. They are known as ETHOLOGICAL ISOLATION. They are expressed in a great variety of different ways. The conspicuous plumage and elaborate, melodious songs of many male birds are among the most familiar examples, although these also

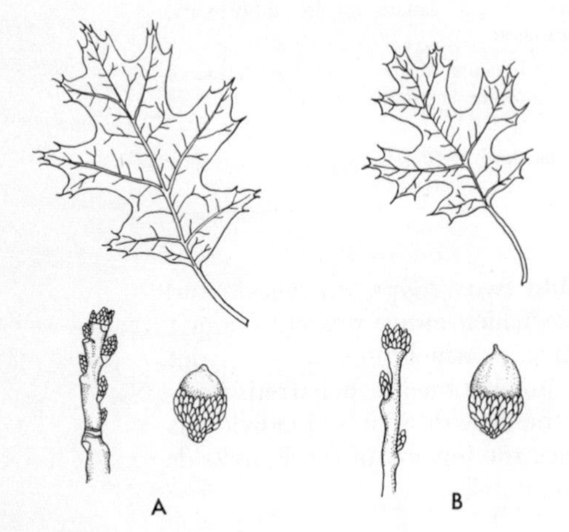

Fig. 5-8. Two species of oak, the black oak (*Quercus velutina*) and scarlet oak (*Q. coccinea*), which in some parts of their sympatric distribution remain quite distinct from each other and in other regions form large intermediate hybrid populations. From Preston, *North American Trees* (Iowa State University Press).

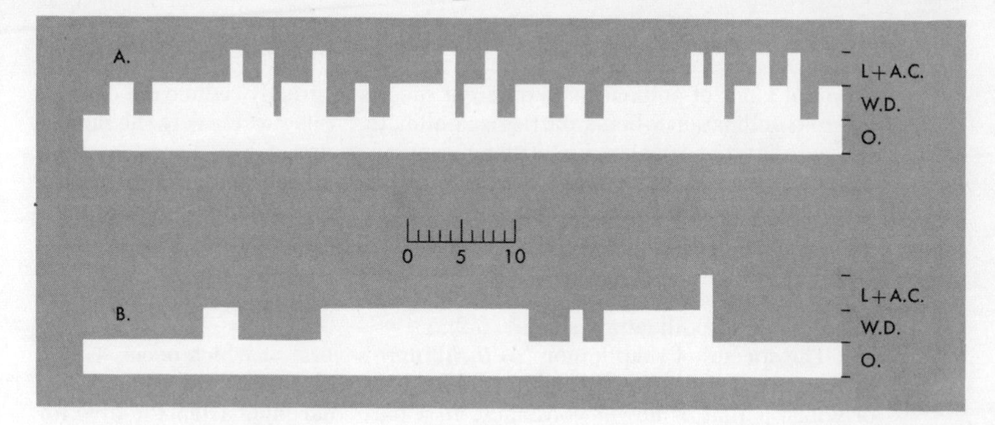

Fig. 5-9. Chart showing the divergent courtship patterns of two closely related species of *Drosophila*. The sequence reads from left to right, on the scale showing time units of $1\frac{1}{2}$ seconds. The height of the black columns indicates the courtship element being performed. The lowest level is O (orientation), the middle level W.D. (wing display), and the highest level L + A.C. (licking and attempted copulation). From Manning, A. in Behaviour, 15:123. 1959.

warn them from approaching. The calls of frogs and toads; the "singing" of crickets, locusts, and katydids; and the flashing of fireflies are also associated with this function.

Even animals which some people might regard as very simple and at a low level of evolution, such as flies, may possess highly complex and specific courtship patterns. In the fly *Drosophila* these have been studied in detail, and experimental evidence has been obtained to show that they act specifically to remove inhibitions. Thus, they render a female susceptible to copulation with males of her own species but not with males of different, closely related species. Males of the related species *Drosophila melanogaster* and *D. simulans* each have a highly specific pattern of courtship movements. These include orienting the body in a particular position facing the female, displaying their wings, and licking the female with their tongues. As shown in Figure 5-9, males of *D. melanogaster* alternate these activities over a period of about a minute in a completely different pattern from that executed by males of *D. simulans*. If females of *D. melanogaster* are placed in a cage with males of *D. simulans*, the courtship patterns of these males do not render them receptive to copulation. The same is true of females of *D. simulans* placed with males of *D. melanogaster*. A number of experiments have shown that this lack of receptivity is due to inhibitions present in the female which are not removed by the wrong type of courtship pattern. These experiments consist of artificially removing the inhibitions of the female. The antennae of females, their most sensitive tactile organs, have been removed, and in other experiments the females have been etherized. After either of these treatments the inhibitions are removed, and males of a different species have no difficulty in effecting copulation.

Mechanical isolation in animals, due to differences in the structure of their genital parts, has in the past been regarded by several zoologists as an important type of barrier. Recent observations and experiments have, however, shown that such differences are relatively ineffective in preventing cross mating between species.

In higher plants, the most effective prezygotic isolating mechanisms are differences in the structure of the flower, which either attract completely different kinds of pollinators to different species or greatly reduce the chances of cross pollination when a particular pollinator visits successively the flowers of two different species. For instance, most species of beard tongue (*Pentstemon*) have blue or purplish flowers, but their size and shape differs greatly from one species to another. Some are adapted to pollination by large bees, such as the carpenter bee, others to bumble bees, still others to small, solitary bees and to wasps. A few species, moreover, have bright red flowers which are tubular in shape, and fit the beaks of hummingbirds, which are their principal pollinators. (Figure 6-2).

The species of snapdragon (*Antirrhinum*), several of which occur wild in Spain, are all pollinated by highly evolved bees with specialized instincts, one of which is that of flower constancy. In a particular flight from the hive for collecting nectar, they visit flowers of only one kind. Experiments in a garden planted with two species of snapdragon have shown that a particular bee will rarely fly from the flower of one species to that of the other, although different bees from the same hive will be visiting both species. Seed taken from these plants produced very few hybrids, although such hybrids could easily be produced by artificial cross pollination.

In other genera, such as the milkweeds (*Asclepias*) and louseworts (*Pedicularis*), flowers of different species are constructed so differently that even if an insect visits successively the flowers of two different species, the pollen is not transferred from the stamens of one species to the stigma of the other. The situation is not unlike that which prevents cross pollination between two similarly constructed flowers of a primrose, as described in Chapter 3. We can emphasize the importance of different flower forms as isolating mechanisms between plant species by saying that a great wealth of modern observation and experimentation has fully verified the conclusion reached by Charles Darwin in his book, *Forms of Flowers*. Darwin concluded that most of the differences between the flowers of related species of plants have evolved in response to the adaptive advantage of cross pollination between individuals of the same species, and the disadvantage of wholesale crossing between members of different species.

Some additional prezygotic isolating mechanisms in both animals and plants act between the time of mating or cross pollination and the actual union of the two gametes. These are, however, most commonly found between species which are relatively distantly related to each other or which are also separated by other kinds of isolating mechanisms. Consequently, they rarely serve as limiting factors in preventing cross fertilization between different species.

POSTZYGOTIC MECHANISMS

Postzygotic isolating mechanisms are expressed in three different ways: the inviability or weakness of the F_1 hybrid itself; the complete or partial sterility of vigorous F_1 hybrids; and the production of many weak or sterile F_2 progeny by vigorous, fertile F_1 hybrids. The action of all of them can be characterized in a general fashion as the inability of the parental genes to work together properly in the cells of the hybrid or its progeny. This is often called GENIC DISHARMONY.

Genic disharmony can be expressed at various developmental stages in different hybrids, but in both plants and animals there are certain stages which are particularly susceptible to it. In animals these are 1. The first cleavage mitosis of the fertilized egg, 2. the period of gastrulation when many genes in the hybrid nucleus are beginning to act for the first time, 3. the formation of the reproductive organs, particularly the testes, 4. the meiotic divisions themselves, and 5. the development of the gametes, after meiosis. Failure at one of the two first stages results in hybrid inviability and at one of the last three, in hybrid sterility. In many instances of hybrid inviability abnormal development begins at gastrulation but continues for some time before the hybrid dies. In some instances hybrids can grow to normal sized adults but possess various kinds of abnormalities. In hybrids of *Drosophila* for instance, the arrangement of the bristles on the body may be irregular and very different from that of both parents.

The causes of genic disharmony are very poorly known. In fact, original research in this field is much needed to help us to understand how different species evolve. Recent experiments suggest that disharmony is associated with the processes of nuclear metabolism, including both DNA replication and the formation of messenger RNA. In one series of experiments, normal zygotic nuclei were removed from the eggs of the leopard frog (*Rana pipiens*), and replaced by corresponding nuclei taken from the wood frog (*R. sylvatica*). The disharmony between nuclei and cytoplasm thus produced caused the embryos to die at the gastrula stage or earlier. More important, however, is the fact that nuclei from these abnormally developing embryos at the middle blastula stage were transplanted back into enucleated eggs of their own species and even then were unable to develop normally. Apparently, interaction between the genes of these nuclei and the cytoplasm of the foreign species had rendered them incapable of directing normal development, even when surrounded by their normal cytoplasmic environment.

Other experiments suggest that genic disharmony is associated with uncoordinated rates of various developmental processes which normally form an integrated sequence. In the leopard frog, hybridization between certain widely different races of the same species, such as those from Vermont and from Florida, produces abnormal, inviable hybrids. Moreover, the abnormalities are quite different depending upon which race is the female parent. If the female parent is the race from Vermont fertilized by sperm of the Florida race, the hybrid tadpoles have greatly enlarged heads and abortive tails; while eggs of the Florida race fertilized by Vermont sperm give larvae with minute, imperfectly developed heads and very large tails. This is associated with the fact that in the Vermont race the embryos develop slowly and have a relatively low temperature optimum, while those of the Florida race develop much more rapidly and optimally at a considerably higher temperature.

Hybrid sterility due to abnormal development of gonads is very common in animals, particularly in the testes of the male. It is the usual condition in viable F_1 hybrids between species of *Drosophila* and other flies; between various species of mammals such as cattle × yak, cattle × buffalo, and horse × zebra; and between some species of birds such as the mallard and muscovy ducks. The abnormality is most strongly expressed as very poor growth and a low rate of mitosis in the cells of the seminal tubules. These normally give rise to the primary spermatocytes in which meiosis occurs. The abnormalities often extend to the processes of meiosis itself, so that even if a few primary spermatocytes are formed, the cells break down and eventually disintegrate without completing the meiotic divisions.

In plants, the developmental stages most susceptible to genic disharmony are in general similar to those in animals. They are 1. the first cleavage mitosis, 2. the early stages of embryo and endosperm development, 3. the germination of seeds and development of the first true leaves, 4. the formation of reproductive structures, particularly the anthers of the stamens, and 5. the development of gametes and gametophytes, both pollen grains and embryo sacs, after meiosis has been completed.

Breakdown in the early stages of embryo development is often due to disharmony between the embryo, the endosperm which provides food material for its growth, and the maternal tissue. In such cases, hybrid embryos can often be obtained by removing the embryo from the seed and culturing it in an artificial nutrient medium. Breakdown at the time of seed germination, like that occurring at the time of gastrulation in animals, is associated with the initial function of a number of genes, particularly those associated with the development of chloroplasts in the leaves. Hybrid weakness in plants is often expressed in the form of chlorosis resulting from poorly developed and few chloroplasts. Vigorous hybrids are sometimes sterile because their reproductive structures, particularly the anthers of their stamens, are abortive and do not develop microspore mother cells, but this condition is less common in plants than in animals.

In both plants and animals hybrid sterility may result from abnormal segregation at meiosis of either whole chromosomes or of blocks of genes contained in chromosomal segments. If the chromosomes of the parental species are strongly differentiated, they cannot pair at all. For instance, in the hybrid between the radish and cabbage, the nine chromosomes derived from the radish may not pair at all with the nine derived from the cabbage, so that at meiosis we see eighteen single chromosomes instead of the nine pairs found in the parental species (Figure 6-6). Since these unpaired chromosomes are unable to line up on the meiotic metaphase spindle, they are distributed irregularly to the poles. The daughter cells thus receive unbalanced complements of chromosomes the genes of which are unable to direct the development of the pollen grains or embryo sacs.

When the parents of a hybrid are more closely related to each other their chromosomes may pair at meiosis, but because the arrangement and structure of genes on the chromosomes is different, their pairing is imperfect and results in the segregation to the gametes of abnormal, disharmonious combinations of genes. An example of this type of behavior is seen in a hybrid between two species of primrose, *P. verticillata* from Arabia and *P. floribunda* from northwestern Pakistan. In this and many other examples, the sterility of the hybrid has been shown to be due to many small differences in structure between the parental chromosomes. When the number of chromosomes in such hybrids is doubled, so that each chromosome becomes present in duplicate, the doubled hybrid is at once fertile and true breeding. This is because pairing occurs between the exactly similar duplicate chromosomes (Figure 6-7). The evolutionary significance in higher plants of such doubled hybrids will be discussed in the next chapter.

In plants and in some lower animals, the evolutionary significance of hybrid sterility which results from differences in chromosomal pattern and consequent abnormal segregation at meiosis is very different from that of hybrid sterility which results from abnormal development of reproductive structures in the F_1 hybrid, and which cannot be corrected by doubling the chromosome number. We can, therefore, designate the former SEGREGATIONAL STERILITY and the latter DEVELOPMENTAL STERILITY. Segregational sterility occurs with about equal frequency in both animals and plants. Since, however, developmental sterility is far more common in animals than in plants and acts at an earlier stage of development, segregational sterility rarely becomes the limiting factor against further reproduction of hybrid animals. In plants, on the other hand, segregational sterility is often the limiting factor. Consequently doubling the chromosome number of interspecific hybrids, which eliminates or greatly reduces segregation of chromosomes and genes derived from different parental species, has given rise to many new species of plants. Some of these have evolved further to give rise to entire genera and subfamilies. This topic will be taken up in the next chapter.

In both animals and plants, there are examples of hybrids which are highly or at least partly fertile, but which give rise to weak, abnormal, or sterile progeny in the second generation. This phenomenon is known as HYBRID BREAKDOWN. From the developmental point of view it is similar to segregational sterility, except that the effects of the segregation of disharmonious gene combinations are delayed until after fertilization.

THE GENETIC BASIS OF ISOLATING MECHANISMS

This extended description of isolating mechanisms has been presented as a prelude to the discussion of what is perhaps the most important single problem of evolution; the origin of reproductively isolated populations or population systems, which can then give rise to new lines of evolution. Before we can begin this discussion, however, we must explore the genetic basis of the different kinds of isolating mechanisms and the relationships between them. We must also learn what connection there is between the factors controlling isolating mechanisms and those responsible for the differences which we see between races, subspecies, and species.

The first important fact in this connection, which has become clear through studies of fertile or partly sterile hybrids, is that each type of isolating mechanism, when well developed, is based upon differences in respect to many different genes or chromosomal segments. In progeny of hybrids between species which differ in respect to habitat preferences, season of reproductive activity, courtship patterns, or flower structure, segregation for these differences never gives clear-cut Mendelian ratios, which would be the case if they were governed by one or a few genes. Instead, we find the complex ratios and the spectrum of intermediate types which we would expect on the basis of multiple factor inheritance. The same is true for both developmental and segregational sterility. Developmental sterility has been studied by T. Dobzhansky in back cross progeny between two species of *Drosophila, D. pseudoobscura* and *D. persimilis*. In the F_1 generation, male hybrids are completely sterile because of much reduced, abnormal testes while females are partly fertile. If, however, the females are back crossed to either parent, the males of the back cross generation show a wide range of testis sizes (Figure 5-10). Since all of the chromosomes in the cultures used of the parental species could be identified by means of recessive genes which they bore, Dr. Dobzhansky was able to determine that the degree of abnormality of the testes of a fly was directly proportional to the number of his autosomes (i.e. chromosomes not directly concerned with differentiation of sex) which were derived from a different species from that which contributed his single X chromosome, belonging to his X-Y pair of sex chromosomes. This shows that genes tending to reduce testis size in the hybrids are found on every chromosome of the parental species.

In some examples of plant hybrids with partial sterility of the segregational type, large F_2 progenies have shown a wide range of fertility, from zero to nearly 100 per cent. This result indicates also that the chromosomal differences responsible for the sterility are numerous, and that each one has a relatively small effect.

In addition, any two species, even those closely related to each other, are usually separated from each other by a number of different kinds of isolating mechanisms. Two species of *Drosophila* which are so much alike

Fig. 5-10. Chart showing the relationship between testis size and chromosomal content in back cross males from the hybrid *Drosophila pseudoobscura* x *persimilis* to its two parental species. Modified from Dobzhansky, *Genetics and the Origin of Species*, 3rd ed. (Columbia University Press).

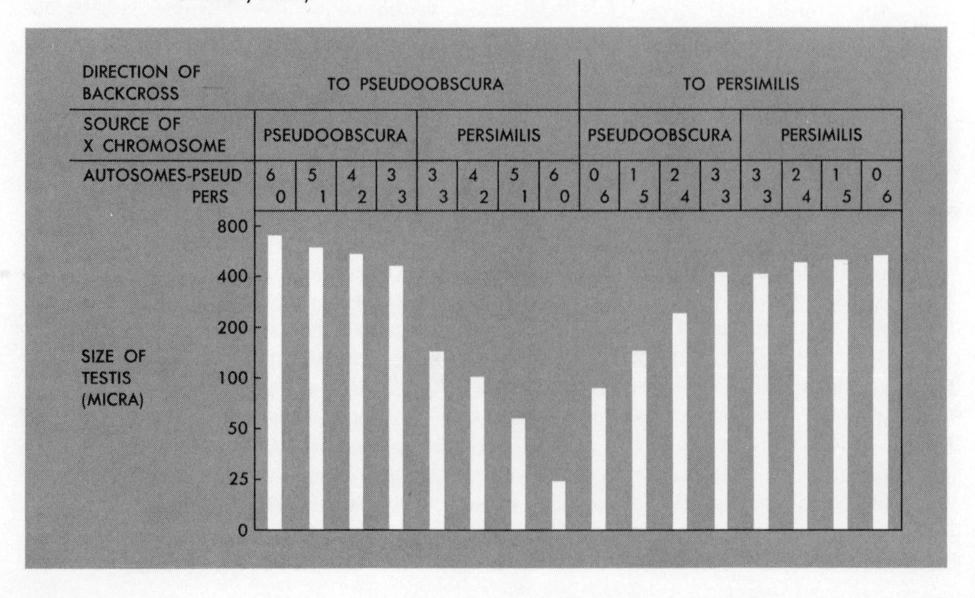

that even experts find them difficult to separate on the basis of the appearance of the adult flies are nevertheless usually separated by a combination of ethological isolation, partial hybrid inviability (many fertilized eggs do not develop), hybrid sterility in the males, and hybrid breakdown of back cross progeny. In plants such as milkweeds, we find differences in habitat preference combined with differences in flower structure, hybrid inviability, and sterility. There are examples of pairs of species which are separated from each other by only one type of isolating mechanism, but these are exceptional.

Taken together, the fact that related species are usually separated from each other by several different isolating mechanisms, plus the fact that each mechanism is determined by several genic or chromosomal differences, tell us that species do not arise at a single step but through the accumulation of many different genetic changes. Still other facts tell us that not all of the differences which contribute to genetic divergence of populations from each other also promote their isolation, but that special kinds of changes are involved.

The first of these is the fact, already brought out at the beginning of this chapter, that populations can diverge to a great extent, both in their outward appearance and their adaptation to different habitats, without becoming separated by any isolating mechanisms. In many instances, these distinctive differences have persisted for millions of years, merely because the populations are separated from each other by long distances, even though they can easily be broken down by artificial hybridization.

SIBLING SPECIES

In other instances we find groups of species which are so much alike that they are almost impossible to distinguish from each other by their outward appearance, but which nevertheless are separated from each other by a number of well-developed isolating mechanisms. These are known as SIBLING SPECIES. A good example is found in the malaria carrying mosquito of Europe (*Anopheles maculipennis*) and its relatives. This is a group of six species which are so much alike that they cannot be told apart on the basis of the adult mosquito. Their larvae or "wrigglers" are, however, somewhat different from each other, and their eggs are so different that officers of the U. S. Army Sanitary Corps, who were in Italy during World War II, found that they could teach an untrained GI to distinguish the different species by their eggs. This recognition is of great practical importance, since some of the species carry malaria and some do not. Consequently the adults do differ from each other in respect to invisible physiological characteristics.

These mosquito species are isolated partly by differences in mating behavior, and partly by hybrid sterility, which is complete in F_1 males of any combination. Consequently, they easily pass the test of sympatry. Populations of two or more species may exist together over very wide areas without the slightest indication of gene exchange between them.

Sibling species are very common in other flies, such as *Drosophila*. Indeed, they occur in all groups of animals. Well-known examples among birds are the tyrant flycatchers, and among fishes the whitefishes in northern

Eurasia and North America. Others have been studied in butterflies, crickets, "water fleas" (*Daphnia*), limpets, and sponges. They are apparently less common in plants, but have been found in the tarweeds (*Holocarpha*) and Gilias of California, and in some genera of grasses (*Elymus, Panicum*).

From the genetic and evolutionary point of view, sibling species are just as valid as are those separated by conspicuous visible characteristics. Their existence tells us, however, that the genetic differences which build up isolating mechanisms can occur independently of those which produce clearly recognizable differences between species. To understand the origin of species, therefore, we must ask ourselves what kinds of evolutionary processes can be particularly responsible for the origin of isolating mechanisms.

THE EVOLUTION OF ISOLATING MECHANISMS

Based upon the large body of evidence which was reviewed in the first four chapters of this book, we now realize that two conditions must be fulfilled before populations can diverge with respect to any characteristic which is determined by many separate genes. First, divergent pressures of natural selection must be acting upon them. Second, they must be well enough isolated from each other so that the initial genetic divergence will not be "swamped" by gene exchange. These conditions must be fulfilled for reproductive isolating mechanisms as much as for any other kind of difference. Consequently, the processes which lead to the origin of species are not of a special kind. They are not qualitatively different from the processes which give rise to divergent races and subspecies. They consist of the divergent action of natural selection upon those particular kinds of differences in genes and chromosomes which can form isolating mechanisms.

The origin of most of the prezygotic isolating mechanisms can be the direct result of selection for divergent adaptations. Habitat isolation can clearly come about through adaptive radiation of populations into the different habitats found in any region. Seasonal isolation can be acquired while populations occur in separate regions with different climates. For instance, among the pines of California there are several examples of pairs of sympatric species which differ markedly in time of pollen shedding. One member of each pair sheds its pollen and fertilizes its cones in February or March, and the other in April or May. In each example, the early flowering species occurs by itself in regions which are farther south, lower in altitude, or nearer the coast than the area of sympatry, and in which the winter climate is relatively mild and dry. The later flowering species, on the other hand, occurs by itself in more northerly regions, or at higher altitudes, where conditions for cross pollination and fertilization are very unfavorable in February and March. Consequently, the most likely evolutionary history of these species pairs is that they were originally allopatric. The later flowering species evolved its distinctive seasonal rhythm in climates which were not favorable for pollination until late spring. When the species later became sympatric in a region having a mild climate, gene combinations for late flowering had been so firmly established that they persisted, even though selection pressures favoring them were not nearly so strong.

In plants, natural selection for flowers adapted to pollination by particular species of insects or other animal vectors could take place where the vectors concerned were unusually abundant, and other vectors were relatively scarce. For instance, most species of larkspur (*Delphinium*) have blue or purple flowers, and are pollinated by bees, whose vision is most acute in the violet and ultraviolet range of the spectrum. In northern California there exists a species with scarlet flowers (*D. nudicaule*). It is pollinated by hummingbirds whose vision is most acute in the red and infrared range of the spectrum. This red larkspur occurs in moist, shady canyons where bees are relatively uncommon and hummingbirds are more abundant. Another flower which is pollinated by hummingbirds, the "California fuchsia" (*Zauschneria*), flowers in late summer and autumn when relatively few native bees are flying. There are, however, many examples of related species having different pollinators which cannot easily be explained on this basis.

The origin by natural selection of ethological isolation is very poorly understood. Weakly developed barriers of this sort are occasionally found between allopatric races of a species and have presumably originated as by-products of selection for adaptation to different environments. Artificial selection in populations of *Drosophila* for differences in bristle number and other characteristics have sometimes led to some degree of ethological isolation between the selected populations. There is, therefore, some reason for believing that ethological isolating mechanisms can be built up as a by-product of divergent selection for adaptive differences, but much more research needs to be done before we can be sure of this.

Genic disharmony in the development of hybrids, which includes both hybrid inviability and developmental hybrid sterility, is most likely to evolve as a by-product of selection for different rates and temperature optima for various processes of cellular metabolism. The beginning of this process is seen in the divergence of the races of the leopard frog, mentioned earlier in this chapter. It has continued to the stage of isolating closely related species in the case of certain Japanese fishes, known as loaches (*Cobitis taenia* et aff.).

The origin of segregational hybrid sterility is best explained by the fact that the various genes which constitute a particular adaptive complex are often grouped together on a small segment of a chromosome, as described in Chapter 3. As populations invade new territories, we might expect these adaptive clusters to become modified by adding or subtracting genes. This could be accomplished most easily by means of small chromosomal rearrangements. Evidence that racial differences of this sort have evolved in plants is provided by detailed studies of chromosome structure in different races of a plant related to the lily family (*Trillium kamchaticum*) in northern Japan. This plant has very large chromosomes which, when exposed to cold temperatures at the time of mitosis, reveal an intricate pattern of differences in staining capacity. Within most populations of Japan, individual differences with respect to some or all of the chromosomes are found, and heterozygosity for different chromosomal types is the rule. Moreover, the different chromosomal types have distinctive geographic distributions, associated with corresponding differences in the outward appearance of the plants. So far as is known, these chromosomal differences are too small to give rise to any lowering of fertility in hybrids between individuals carrying different chromosomes. Nevertheless, chromosomal differences between *T. kamchaticum* and other species of *Trillium,* which are great enough to make their hybrids sterile, are greater in quantity but are otherwise similar to the interracial differences within *T. kamchaticum.*

Chromosomal differences of this sort may in some instances arise in rapid succession, in response to rapid, drastic alterations in the environment. Among the species of *Clarkia,* a genus of annual flowering plants found in California, there are several species which are found only in a very restricted area, and which are closely related to more widespread species occurring in nearby regions. When crosses were made between these localized species and their more widespread neighbors, the F_1 hybrids in every case proved to be heterozygous for many chromosomal differences and were consequently sterile. The places where these species occur are known to have been subjected to drastic changes in climate in recent times. Consequently, a likely explanation of their origin is that they were derived from their neighbors via a rapid succession of chromosomal changes, which brought together groups of genes adapting them to their new environments.

It is not hard for us to see that the accumulation of numerous genetic differences constituting the system of isolating mechanisms that separate any two related species could not arise if gene exchange were occurring even at a low rate between the populations concerned. Such exchange would act directly to break down the isolation, while the selective pressure toward maintaining and increasing it would be indirect in nature. Consequently, those isolating mechanisms which enable two populations eventually to exist side by side and pass the test of sympatry must arise while they are separated from each other to such a degree that gene exchange between them is impossible. The degree of separation needed is still not clear, and probably differs according to the organism involved. Organisms which easily travel for long distances, such as birds and the larger mammals, probably need to be separated in different geographic regions before isolating mechanisms between them can be built up. The same is true of the trees of the temperate zone, which have pollen and seeds that can be borne for long distances either by the wind or by strong flying birds. For small, sedentary organisms, and particularly for specialized parasites, the amount of spatial isolation needed is much less.

If we assume that geographic separation is usually needed before reproductive isolation and the origin of distinct species can take place, then we might expect to find a characteristic distributional pattern in groups containing species which are actively evolving. Populations between which isolating mechanisms are partly developed to varying degrees should be separated geographically in regions either close to each other or having somewhat similar environmental conditions. Distinct, sympatric species should be more strongly isolated from and more different from each other than are the related allopatric populations.

Just such a pattern has been found in plants of the daisy family known as "tidy tips" (*Layia*) in central California. These populations have been carefully analyzed in respect to the morphological differences between them, their geographical distribution and habitat preferences, and the degree of sterility and of chromosomal differentiation found in hybrids between them.

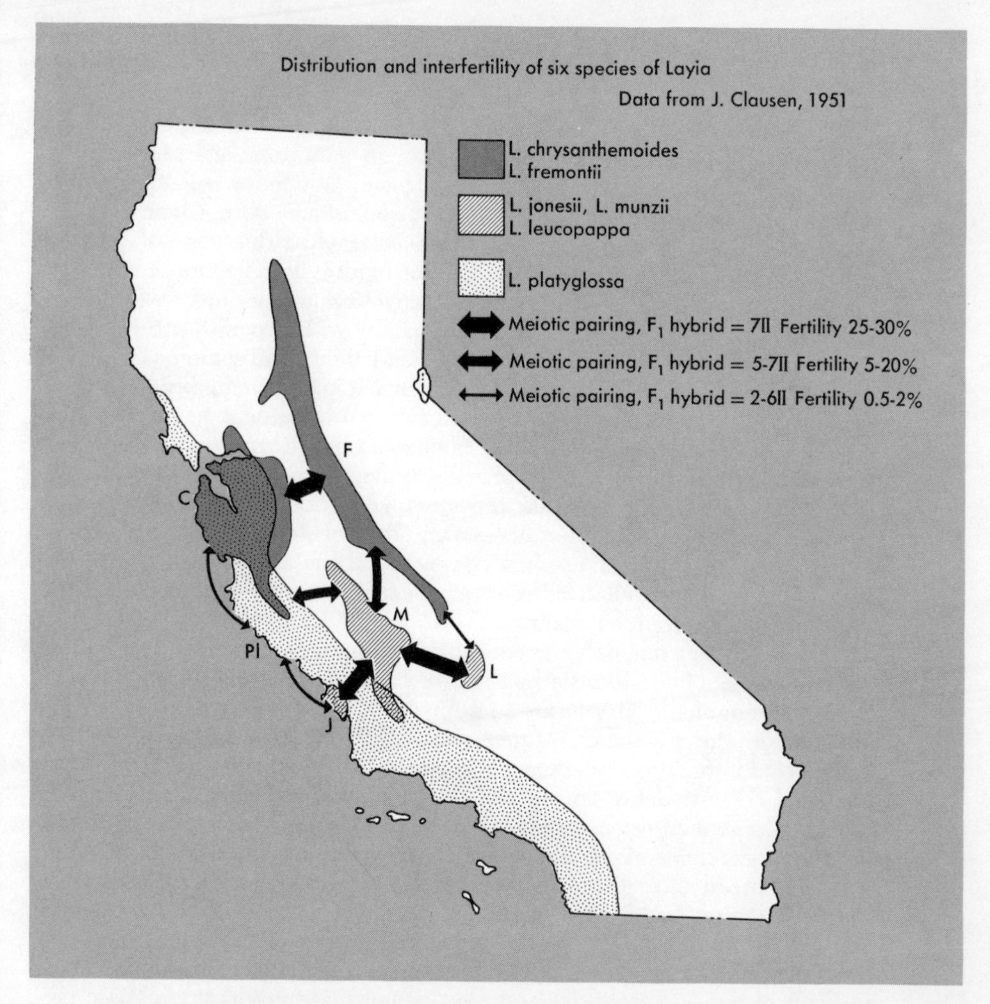

Fig. 5-11. Distribution and fertility relationships of six species of the genus *Layia,* or "tidy tips" (Compositae), found in California. Further explanation in the text.

The relationships between six closely related species of this genus are shown in Figure 5-11. On the basis of geographic distribution they fall into three groups. One group consists of a pair of species, one in the region about San Francisco Bay, and the other in the foothills of the Sierra Nevada directly east of there. The second group contains three species which occur about 200 miles south of San Francisco, one on the coast, one about 50 miles inland, and the third about 100 miles inland. The sixth occurs by itself in southern California, but extends northward so as to be partly sympatric with members of both of the two other groups.

When these species are artificially intercrossed, vigorous hybrids can be formed between all of them, indicating that they are all fairly closely related to each other. Hybrids between the two species within the first group and the three within the second group have about 25 to 50 per cent of normal fertility. This indicates that these populations are on the borderline between subspecies and species. One could not be sure what would happen if they should occur naturally in the same region; they might pass the test of sympatry or they might become merged through gene exchange. Hybrids between populations belonging to the first group and those belonging to the second are much more sterile, and their chromosomes do not pair normally at meiosis. These two groups have certainly reached the stage of separate species. Finally the single species in the third group forms almost completely sterile hybrids with all of the others, and its chromosomes are well differentiated from theirs. Where it occurs sympatrically with two of the other species it occupies completely different habitats, and there is absolutely no sign of gene exchange between them. Here we can see in diagrammatic form three stages in the process termed by evolutionists GEOGRAPHIC SPECIATION.

The discussion presented in this section can be summarized as follows. If two populations become geographically isolated from each other, and are subjected to divergent selection pressures, an incidental byproduct of this selection may be, although not necessarily, the development of various kinds of reproductive isolating mechanisms. Once these have become well developed, the two populations rank as species, and can become sympatric again without losing their identity.

Under these conditions of secondary sympatry, reproductive isolating mechanisms can be reinforced by the *direct* action of natural selection on the barriers themselves. This is accomplished by the direct action of natural selection in the following manner. Individuals which tend to mate with members of a foreign population will produce fewer vigorous, fertile offspring than will those which mate exclusively with other individuals of their own species. Consequently, mutations or gene combinations which favor mating between individuals of the same species will have a positive selective value, and will spread through the population. The effectiveness of such selection has been shown by an experiment in which populations of two species of *Drosophila* separated by a partly developed barrier of ethological isolation were placed together in a cage. The experimenter then removed systematically all of the F_1 hybrids formed, thus penalizing reproductively those flies which mated with the other species. After sixteen generations flies of the two species were found to be almost completely isolated from each other.

These results show that ethological isolation can be reinforced by natural selection when two sympatric species often come in contact with each other. We might expect natural selection to reinforce similarly any isolating mechanism which tends to prevent hybridization at relatively early stages of the process. For instance, if two sympatric species often form sterile hybrids which automatically reduce the reproductive potential of both of the parental species, genetic changes which in any way prevent the hybrid zygotes from being formed will raise the reproductive potential and hence increase the evolutionary success of these species. In animals this can be done by reinforcing ethological isolation, and by various mechanisms which prevent fertilization even though copulation has taken place. In plants, natural selection might be expected to favor increasing divergence in flower structure between sympatric species. It would favor genes which prevent pollen of a foreign species from germinating on the stigma of a species, or would cause foreign pollen tubes to grow more slowly than those of the same species. In several groups of flowering plants, such as the milkweeds (*Asclepias*), Gilias, and ground-cherries (*Physalis*), barriers which block the initial stages of hybridization are strongly developed between sympatric species.

Based upon the concept of geographic speciation followed by reinforcement of isolating mechanisms, we should expect to find different relationships between species inhabiting unstable regions as compared to those inhabiting regions with more stable environments. In regions where the climate is changing relatively rapidly in terms of the geological or evolutionary time scale, or in which mountain building movements and volcanic activity are taking place, areas having a particular set of environmental conditions are changing their position rapidly, and the organisms inhabiting these areas must change their geographical distributions accordingly. In such regions, we should therefore expect to find many examples of allopatric populations on the borderline between races and species, as well as many species which have only recently become sympatric and are separated by relatively poorly developed isolating mechanisms. On the other hand, regions which have been climatically and geologically stable for long periods of time possess environmental niches which have remained the same for corresponding lengths of time. In such regions, we would expect to find fewer examples of species which have recently altered their geographic ranges. The opportunities for geographic speciation would be less. We should, however, expect to find more examples of species which have existed side by side in the same region for very long periods of time, and which have consequently undergone a long period of reinforcement of reproductive isolation by natural selection. These species should, therefore, be sharply distinct and well isolated from each other.

A comprehensive review of hybridization in fishes has revealed exactly this situation. In regions like the streams flowing into the desert from the mountains of Southern California, which vary greatly from season to season according to the amount of rainfall, and the relatively new, unstable lakes in the glaciated regions of northern Europe, the fish fauna is poor in species, and the species are poorly defined. Hybridization between species is common. The same is true of the lakes and streams of Scandinavia and Canada, which were formed after the retreat of the glaciers a few thousand years ago. On the other hand, the coral reefs of tropical regions are an old habitat which has remained stable for a very long period of time. There the fish fauna is much richer, and includes many groups of related species which are sympatric over wide areas, but are everywhere very distinct and strongly isolated from each other. This situation is further evidence for the theory that rapid evolution is associated with habitat disturbance, and evolutionary stabilization is the result of long continued stability of the environment.

Chapter Summary

Most species of higher animals and plants contain many populations which differ from each other in adaptive properties and visible characteristics, but which intergrade in intermediate environments. These are generally known as races, and those most easily recognized are called subspecies. Races and subspecies form completely fertile hybrids when intercrossed, and differ from each other in respect to a large number of genes. Under certain conditions, they can remain as different races for millions of years without evolving to the stage of distinct species.

Different species can be most easily recognized when they exist together in the same area, i.e. are sympatric, and maintain their identity because they do not exchange genes with each other or do so only to a limited extent. This is because they are separated from each other by various kinds of ISOLATING MECHANISMS. These may either prevent hybrids from being formed in nature or may consist of the inviability or sterility of interspecific hybrids.

Even closely related species are usually separated from each other by several different kinds of isolating mechanisms, and each mechanism is based upon the action of many different genes. This shows that effective isolating mechanisms evolve through the accumulation of many genetic differences of particular types. In some instances, these differences have little effect on the outward appearance of the organism, so that strongly isolated species can be very much alike in appearance. These are called SIBLING SPECIES.

The origin of isolating mechanisms is by mutation, genetic recombination, and natural selection of particular kinds of differences. When two sympatric species are strongly but not completely isolated from each other, natural selection can act to strengthen these mechanisms. Divergence in respect to reproductive isolating mechanisms is very unlikely to begin unless populations are initially isolated from each other spatially at least to some degree. In regions with unstable, rapidly changing habitats, isolating mechanisms are usually more poorly developed than in regions where the environment has remained stable for long periods of time.

Questions for Thought and Discussion

1. Does the evidence from reproductive isolating mechanisms as they exist in nature favor the hypothesis that species have arisen at a single step? Use specific examples to explain your answer.

2. Using examples, contrast sibling species with polytypic species, and explain the bearing of this contrast on the relationship between evolutionary divergence and the origin of species.

3. In what ways can natural selection act to promote the establishment of reproductive isolating mechanisms?

4. Why, on theoretical grounds, is spatial or geographic isolation between populations considered by many evolutionists to be necessary before these populations can become separated by reproductive isolating mechanisms? What factual evidence indicates that this is actually the case?

CHAPTER 6

The role of hybridization in evolution

In the last chapter, we learned that biologically distinct, reproductively isolated species are most likely to evolve in an environment which is unstable in terms of the evolutionary time scale, that is, over periods of hundreds or thousands of years. In such an environment, the conditions which favor populations having a particular set of adaptive characteristics will be constantly shifting back and forth from one region to another. That the populations of plants and animals themselves will also be shifting about in this way is a foregone conclusion.

Given these unstable conditions, we would expect to find many examples of previously separated populations coming together and having opportunities for their individuals to hybridize with each other. The evolutionary consequences of such hybridization will be explored in the present chapter.

EVOLUTIONARY DEFINITION AND IMMEDIATE EFFECTS OF HYBRIDIZATION Biologists have used the term hybridization in many different ways. Geneticists apply the term to crossing between pure lines which differ by a single gene, as in the experiments of Gregor Mendel. Plant breeders speak of hybrid corn which is the result of crossing inbred lines differing in respect to many genes. We also speak of hybridization between breeds or races of domestic animals, and between species, such as the horse and donkey. If the evolutionist is to use the term hybridization, he must define it in a way which will be reasonably precise and meaningful to him. Since evolution consists chiefly of changes in adaptive complexes of genes, the most meaningful definition of hybridization to an evolutionist is crossing between populations having different adaptive gene complexes. Such populations may be either different races or subspecies of the same species, or different species which are separated by variously developed isolating mechanisms.

Hybridization between populations having different adaptive gene combinations can increase greatly the size of the gene pool with respect to genes having different adaptive values, provided that the hybrids can give rise to segregating progeny in later generations. Most of the segregates produced from such hybrids will, of course, have a lower adaptive value than that of either of the parental populations, particularly in the environments to which these parental populations are adapted. A small proportion of them may, however, prove to be better adapted to certain environments. Natural hybridization between differently adapted wild races is in many ways comparable to the artificial hybridization carried out by animal and plant breeders with widely different varieties of domestic animals or cultivated plants. If, for instance, we cross two widely different varieties of *Iris* or *Gladiolus,* we get in the second generation an enormous array of types, having every conceivable combination of the characters of plant size, earliness, flower size, texture of petals, and flower color which were present in the parents, plus many characteristics which appear to be new. Most of these progeny are in many ways less vigorous or less attractive than their parents. The skilled breeder can, nevertheless, select from this mass of variants one out of a hundred or even one out of a thousand individuals which in various ways is superior to either parent. Similarly, even though the great majority of the progeny of hybrids between two natural varieties or species are poorer in every respect than their parents, the forces of natural selection are strong enough to pick out that tiny fraction of them which may be better adapted to an available environment, and to increase greatly the frequency of such gene combinations in the environment to which they are adapted.

There are, however, two drawbacks to the success of such hybrids. In the first place, most hybrids between species are wholly or partly sterile, so that segregating offspring from them are difficult or impossible to obtain. Secondly, the genotypes found even in the second generation of a wide cross are highly heterozygous, and so will not breed true. Hybridization will, therefore, be an important factor in evolution only if the three following conditions are met. 1. There must be one or more environmental niches available to which certain hybrid derivatives are better adapted than are either of the parental populations. 2. In the progeny of partly sterile hybrids, fertility must be restored while valuable new gene combinations resulting from hybridization are still retained. 3. New combinations must become sufficiently constant genetically so that they can maintain themselves under natural conditions.

When these three factors are taken together, they make us realize that the successful contributions of hybrid progeny to evolution will always be rare, and in many groups will be highly exceptional events. Evolution, however, does not always progress through gradual accumulation of the kinds of changes which occur most commonly in populations. At many times during the course of evolution, combinations of events which occur with a very low frequency have effects far more significant than might be imagined on the basis of their infrequency. With this consideration in mind, ways will be considered by which populations can overcome these three restrictions or limitations to the success of hybrid progeny.

HYBRIDIZATION IN OLD AND NEW ENVIRONMENTS

Anyone familiar with animal and plant communities in nature realizes the fact that if a particular set of environmental conditions have existed for a long time, every available environmental niche is filled with well-adapted species of animals and plants, so that populations with new adaptive characteristics have "no place to go." On the other hand, when such communities are disturbed in any way, the equilibrium between their species is upset, so that new kinds of plants and animals have a chance to enter them. This is most likely to be true if the disturbance is accompanied by a change in climate.

In our modern world, the commonest cause of such disturbances is man himself. We are hewing down forests; ploughing and bulldozing land; destroying natural predators; introducing domestic animals, which devour vegetation much faster and more indiscriminatingly than do the native animals; and spreading weeds, pests, and diseases at a rate which is many times faster than any which the world has seen in its long history. The destructive effects of these disturbances are familiar to everyone and are the subject of numerous books and other polemics. Since construction is always slower than destruction, the evolutionary changes which these disturbances have promoted are much more subtle and hard to recognize. A small-scale example of them is industrial melanism, which was discussed in Chapter 4. If we study the wild animals and plants which have been able to profit from man's disturbances, we find many more extensive examples. At least in plants many of these involve hybridization.

When the early colonists came to North America from Europe, they cut down many of the forests and made open hillsides and cow pastures. Among the plants which took advantage of this change were the hawthorns. These shrubs, which are sun lovers, were originally confined to stream banks, open glades, rocky hillsides, and similar gaps in the otherwise continuous primeval forest. Thus, they not only were uncommon, but also were separated from each other in different kinds of habitats. A number of distinct, sympatric species were present, but we cannot now be sure how many. As pastures were made hawthorns invaded them, and since their seedlings were protected from grazing cattle by their strong spines, they were able to spread rapidly. Consequently, species which had previously been separated from each other by ecological isolation were able to come together. They created swarms of hybrids which are now so complex that in spite of repeated attempts, no botanist has been able to classify them in any reasonable fashion. Some of these hybrids are fertile when cross pollinated, while others have evolved ways of producing seeds by asexual means.

Many groups of weeds, particularly in the grass family, contain species which on the basis of cytogenetic evidence are clearly of recent origin and are found only in habitats disturbed by man. A genus related to wheat and known as goat grass (*Aegilops*), has been carefully analyzed by several geneticists. Its species originated in the Middle East, where man first cultivated wheat and barley about 10,000 years ago. Several of these species have

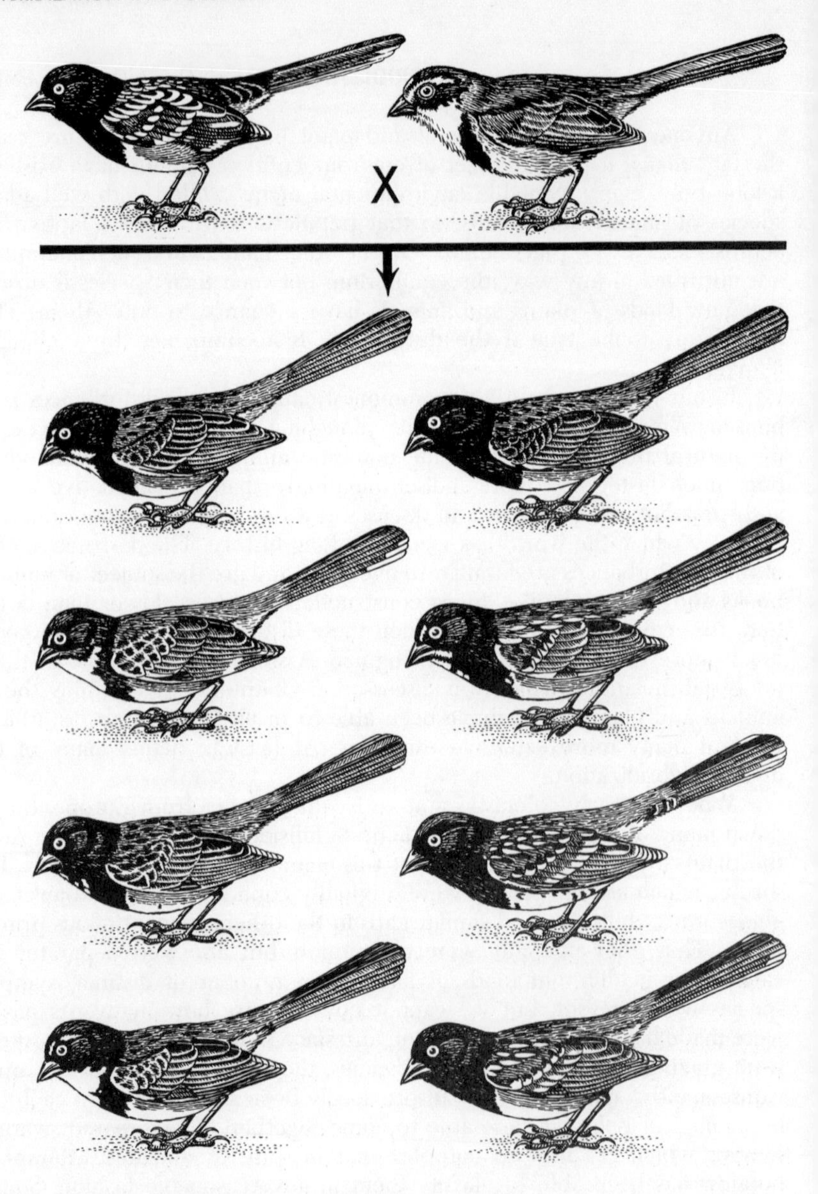

almost certainly evolved since that time, and some of them are clearly of hybrid origin. A similar origin is equally probable for a wild relative of corn, teosinte, which grows in Mexico and Central America.

The evolutionary history of the animals which have invaded the new habitats created by man is harder to follow. Nevertheless, a good deal of evidence indicates that many new subspecies have evolved as a result of this activity, while some species may have evolved in response to it. The house sparrow, or "English sparrow," as it is called in most of North America,

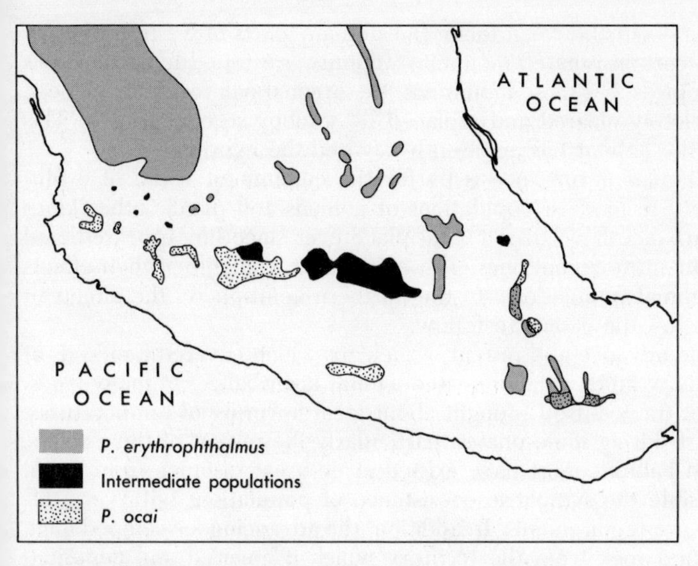

Fig. 6-1. Two species of towhee, *Pipilo erythrophthalmus* and *P. ocai*, together with their hybrids, and a map of their distribution in central Mexico. In the eastern part of this region they occur together with little or no hybridization. In the western and central areas, however, intermediate and strongly segregating populations (indicated in black) are found in many places. From Sibley, University of California Publications in Zoology, 50:109–194, 1950; and Evolution, 8:255, 1954.

arrived here with the early colonists, and is now one of our most common birds. While most of the American house sparrows are indistinguishable from their English ancestors, subspecies have evolved when this bird has invaded areas such as the arid southwest and Hawaii, in which the climate and vegetation are very different from that to which the species was adjusted when it evolved in its original European home. For various reasons, the origin of species in connection with human disturbances in prehistoric times is harder to trace in animals than it is in plants. There are, however, several species which are now found only in habitats associated with man or created by him, and which may therefore have evolved in response to his activity. Among them are the house mouse and the species of *Drosophila* most familiar to geneticists, *D. melanogaster*.

Although the animal examples given above are not known to involve hybridization, and in the case of the house sparrow certainly do not, several examples are known of animal species which are sharply distinct from each other in undisturbed areas, but which have formed hybrid swarms in habitats disturbed by man.

One of the best of these is that of the spotted towhee and collared towhee in Mexico (Figure 6-1). The spotted towhee, a bird familiar to people living in our southern states and California, has a black and white head and back and rufous flanks. The related collared towhee, found in the mountains of Mexico, is mostly green and brownish olive. In the mountains of southeastern Mexico the two species are found together in their typical form, and little or no hybridization occurs between them. But in many parts of western Mexico, particularly on certain isolated mountain summits, are found hybrid swarms of intermediate birds. On these mountains, the original oak forests have been partly or completely cleared and replaced by scrubby second growth. This disturbance of the habitat has apparently favored the hybrids.

Although human disturbance is by far the commonest cause of evolutionary instability in modern populations of animals and plants, other kinds of habitat disturbance have been taking place ever since life appeared, and in the case of the more recent ones clear evidence is available of their effects on plant and animal populations. In the northern hemisphere, the effects of the last ice age are the easiest to follow.

The fourfold advance and retreat of ice caps which covered hundreds of thousands of square miles of land created habitat disturbances of many kinds. The presence of the ice itself brought about local lowering of temperatures, although at least during some phases, particularly the retreat of the glaciers, relatively warm habitats must have existed at no great distance from them. This made possible the sympatric coexistence of populations with very different temperature requirements. In addition, the advancing ice scraped huge masses of soil and rock from the territory which it covered and deposited them along its southern margins, thereby creating a whole series of completely new habitats. Finally, the retreat of the glaciers laid bare tens of thousands of square miles of territory in which the earth's surface had been so profoundly altered that it consisted entirely of a vast mosaic of new habitats. These included not only terrestrial habitats with greatly altered soil conditions, but also innumerable new aquatic habitats in the form of lakes and swamps which were formed from the basins scooped out by the ice or dammed up by masses of glacial debris.

Many groups of animals and plants are known to have evolved in response to these changes. In some groups of mammals this evolution is documented by the fossil record. For instance, several species of bears and of bisons appeared and became extinct during the glacial epoch, as did the well-known woolly mammoth. In plants, active evolution of races and species in the glaciated areas is found in many groups. Two of the most conspicuous examples are willows and birches. In eastern North America, the species of birch trees found south of the former glacial margin are all clearly distinct from each other. They occupy for the most part stable forest communities and are apparently ancient. In the glaciated areas populations of birch trees, particularly those related to the white or canoe birch, are extremely variable and hard to classify into species, indicating that they are unstabilized and still evolving. The same is true of populations of birches in northern Eurasia, from Siberia west to Scotland. Evidence from both phenotypic variability and chromosomal numbers and behavior indicates that their evolution is associated with extensive hybridization.

A good example in fishes is the group of whitefishes and ciscoes found in lakes, streams, and occasionally shallow seas in northern Eurasia and North America. These fishes are mostly confined to drainage basins which were greatly altered by the ice age. Species have been particularly hard to recognize in the group, and different authorities have interpreted them rather differently. The most careful recent studies have been carried out by Swedish specialists, and have been supplemented by artificial hybridization and observations on progeny reared in hatcheries under controlled conditions. These studies have shown that many lakes and streams contain two or three species which are very hard to tell apart, but which can nevertheless be recognized as distinct when carefully studied, and pass the test of sympatry in most places where they occur together. Furthermore, each of these species is found in several lakes over a rather wide area. Some of the lakes, however, contain populations intermediate between these widespread and otherwise distinct species. These intermediate populations are probably hybrid swarms. Apparently, variability has been considerably increased by hybridization between species in some of the lakes which these fishes have occupied since the glaciers retreated and made these habitats available.

In addition to human disturbance and glaciation, several other kinds of environmental upheavals which have occurred in geologically recent times have almost certainly stimulated evolution in the regions where they occurred. In plant populations this evolution appears to have been accompanied by hybridization. In the southwest, from Texas to southern California, the advance of deserts accompanied by earthquakes and other mountain building movements have had such effects. In Florida, the advance and retreat of shallow seas have been chiefly responsible. In Hawaii, entire islands have been built up by volcanic activity and have wholly or partly subsided into the sea. On individual islands, opportunities for evolution (including hybridization) have been created by recent volcanic eruptions. These eruptions have built up new craters consisting of raw lava, cinders, and unweathered soil, side by side with ancient, mature volcanic slopes. Biologists desiring to find examples of evolution in action have learned to look for them among the populations found in all of these different kinds of disturbed areas. They have rarely been disappointed.

STABILIZATION OF HYBRID PROGENY

The two other conditions which hybrid progeny must meet in order to achieve evolutionary success, i.e. restoration of fertility and stabilization of intermediate progeny, will be considered together. From the genetical point of view they both involve the following principle. If a particular difference between two genotypes is governed by several pairs of non-allelic genes, then an intermediate condition with respect to the character can be achieved in two different ways: 1. heterozygosity at all of the gene loci concerned, or 2. homozygosity at all of them, but in a different combination from that possessed by either parent. For example, if the parents differ with respect to four gene loci controlling the same character, and are homozygous at each of these loci, their respective genotypes may be written as follows: *AABBCCDD* and *aabbccdd*. (In this case the use of capitals versus small letters is merely for convenience, and does not indicate dominance or recessiveness.) The intermediate F_1 hybrid would then have the constitution *AaBbCcDd*. This hybrid would, of course, segregate toward both parental conditions at each locus. These segregants, however, would include six homozygotes which would have an equal contribution of genes from both parents, and would, consequently breed true for the intermediate condition: *AABBccdd, AAbbCCdd, AAbbccDD, aaBBCCdd, aaBBccDD*, and *aabbCCDD*. These new, intermediate homozygotes would, therefore, outnumber by threefold the two parental homozygotes. Since they would certainly be formed, the critical question is: Would they be adapted to any of the new environments which might become available?

This question must be answered differently depending upon the character concerned. For differences in adaptation to the external environment the answer is often positive, provided that enough new habitats are available. Among higher plants, several examples exist of intermediate subspecies which occupy habitats intermediate between and more recent than those of more widespread, probably ancestral subspecies. In the Sierra Nevada of California many of the meadows, which were formed after the retreat of the glaciers, are occupied by a subspecies of the sticky cinquefoil (*Potentilla glandulosa*) which is intermediate between two more widespread subspecies. One of these subspecies inhabits warm, dry slopes below the former margin of the glaciers, and the other is found at higher altitudes in both glaciated and non-glaciated parts of the mountains. Among birds, certain oases in the northern part of the Sahara desert are occupied by populations of sparrows which apparently arose through hybridization between the house sparrow and the willow sparrow, a species native to the Mediterranean region.

In respect to those genetic differences which form the principal isolating mechanisms between species of animals: ethological isolation, hybrid weakness, and developmental hybrid sterility, successful intermediate combinations derived from hybridization are not known. One might expect them to be rare, since the genes which cooperate to produce these barriers form complex, integrated systems. Consequently, most of the recombinations between them would be badly maladjusted within themselves, regardless of the external environment. Successful recombinations of the genes controlling such barriers must not, however, be regarded as impossible. Experiments to find out whether they can occur need to be performed.

Among plants, recombination of the genes which control isolating mechanisms can produce new genotypes which are partly isolated from both of their original parents. This has been clearly demonstrated in respect to both floral isolation and segregational hybrid sterility. In respect to floral isolation, good examples can be found in the beard tongues (*Pentstemon*) of the western United States. This genus contains a large number of species which are distinguished from each other principally in the shapes, sizes, and colors of their flowers. Four of the best known species in southern California are shown in Figure 6-2. One of them, found in pine forests in the mountains, has large, two-lipped, widely gaping corollas which are bright blue in color. A second, known as scarlet bugler, which is widespread in the drier parts of southern California, has scarlet corollas, tubular in shape. A third species, the showy pentstemon, lives on brush-covered hills, a relatively recent habitat which is intermediate between those of the mountain pentstemon and the scarlet bugler. It has blue or slightly purplish flowers, intermediate in shape between the other two. In a few places, natural hybrids between the mountain pentstemon and the scarlet bugler have been found, and these are much like the showy pentstemon in a number of respects. The three species are

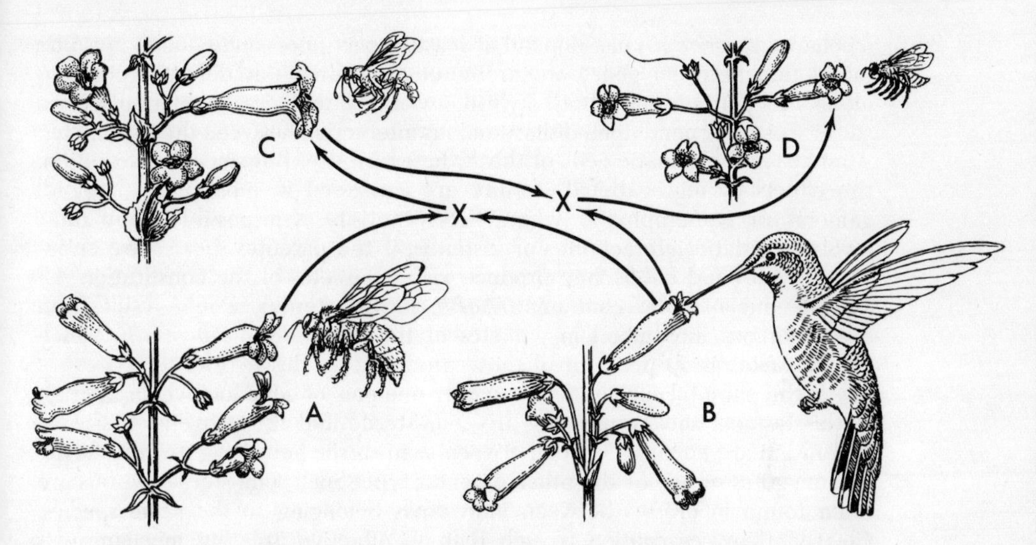

Fig. 6-2. Four species of the genus *Pentstemon* found in California, together with their pollinators. Distributional evidence indicates that the two oldest species are the extreme types, mountain *Pentstemon* (*P. Grinnellii*) (A) and scarlet bugler (*P. centranthifolius*) (B). The two other species, as indicated by the arrows, were derived by hybridization followed by isolation between these two, followed by isolation and stabilization of intermediate populations through the selective action of specific pollinators.

normally visited by the following pollinators: the mountain pentstemon by large carpenter bees (*Xylocopa*); the scarlet bugler by hummingbirds; and the showy pentstemon by solitary wasps (*Pseudomasaris*).

We can reconstruct a probable sequence of events which gave rise to the showy pentstemon as follows. At one time in the past, the mountains of southern California were not brush covered, but were clothed with forests except for the steeper, drier slopes, which were relatively bare. Consequently only habitats suitable to the mountain pentstemon and the scarlet bugler were available. At that time, hybridization between the two species may have occurred from time to time, but was of little consequence because no site was available where populations of hybrid origin could become established. As the climate in the mountains became drier, and forests were replaced by brush, such habitats became available. Some pentstemons of hybrid origin moved into these habitats and in one or more places began to be visited by wasps. As a consequence, natural selection brought about stabilization of gene combinations adapting the vegetative parts of the plant to life on the dry, brush-covered slopes and the flowers to pollination by the wasps which were their commonest visitors in their new habitat. A similar course of events may have given rise to a large proportion of the diverse species of *Pentstemon* found in our western states. Artificial hybridizations between many pairs of related species have usually produced fertile progeny, so that the present barriers which separate them are largely ecological or ecogeographic isolation plus floral isolation.

In respect to segregational hybrid sterility, the genetic mechanisms controlling this barrier are more complex than those controlling other isolating mechanisms, since it must depend at least in part upon unfavorable epistatic interaction between genes or chromosomal segments at different positions along the chromosome. Both hybrid inviability and developmental hybrid sterility often depend upon disharmonious interaction between the two alleles, *A* and *a*, in the somatic cells of the F_1 heterozygote. But since, by definition, the effects of segregational sterility are expressed in monoploid (haploid) gametes or gametophytes, where heterozygosity is impossible, they must involve epistatic interaction. For instance, if the parental species are designated *AABB* and *aabb*, they produce viable gametes of the constitution *AB* and *ab* and a hybrid containing *AaBb*. Partial sterility would result from disharmonious interaction in gametes of the constitution *Ab* or *aB*, which would make up 50 per cent of those produced by this hybrid.

If the parental species differ by only one pair of gene loci which interact in this fashion, the F_1 progeny are semi-sterile and the progeny of the F_2 generation are equally divided between semi-sterile heterozygotes and fertile homozygotes of one or the other parental type. Such semi-sterile hybrids are often found in crosses between individuals belonging to the same species. Clearly, their segregation is such that no effective isolating mechanism is present. If, however, the parents differ by several such pairs of complementary gene loci, the proportion of viable gametes formed by their F_1 hybrid is $\frac{1}{2}^n$, where n is the number of pairs involved. Consequently, the effectiveness of segregational sterility depends upon the number of independently segregating units of which it consists.

Because of this fact, selection for increased fertility in the progeny of a hybrid characterized by this type of sterility has a very high probability of giving rise to a stabilized type which has a different genetic constitution than that found in either parent, and which forms sterile hybrids in crosses with

Fig. 6-3. Diagram showing how segregation from a partly fertile hybrid between two species which differ from each other with respect to many factors affecting fertility can give rise to stabilized, fertile derivatives which form partly sterile hybrids when crossed back to either of their original parents.

Parental Genotypes	AABBCCDD	X	aabbccdd.
Gametes (All viable)	ABCD		abcd.
F_1 Genotype	AaBbCcDd		
Possible Gametes.	Inviable: AbCD , AbCd , Abcd, aBCD ,aBCd , aBcd, ABCd , abCd , ABcD abcD ,AbcD , aBcD . Fertile: ABCD , ABcd , abCD , abcd.		
Possible homozygous and fertile genotypes in F_2.	AABBCCDD	aabbccdd	
Since hybrids between any two of these four genotypes would be partly sterile, the last two could form the starting point for new species.	AABBccdd	aabbCCDD	

both of them. This fact can be illustrated by the simplest possible model, which consists of two independently acting pairs of loci (Figure 6-3). As already pointed out, the semi-sterility resulting from the cross diagrammed in this figure would not be an effective barrier to gene exchange. If, however, the parental species should differ by several pairs of complementary gene loci, fertile genotypes could be derived from them which would form much more sterile hybrids in back crosses with both parents.

If the parents differed by n pairs of loci, the proportion of homozygous, fertile types in later generations, as well as that of viable gametes formed by the F_1 hybrid, would be $\frac{1}{2}^n$. Among this proportion, however, $1 - \frac{1}{2}^{n-1}$ would be new combinations, forming partly sterile hybrids when crossed back to both of the parental species. If $n = 6$, this proportion becomes $\frac{31}{32}$, or about 97 per cent. These figures show that if an effective barrier of segregational hybrid sterility is present, selection for increased fertility in later generations can occur. If it does, the resulting fertile types are almost certain to be at least partly isolated from both of the parental species and so to be potential progenitors of a new species. Since, however, selection for fertility from such hybrids would take place most easily if they were repeatedly back crossed to one of their parents, these new combinations would be likely to arise only if the species were normally self-fertilizing, or if the hybrids had migrated into a habitat not occupied by either of the parental species.

A result similar to that which would be predicted on the basis of these calculations has been obtained in progeny of an artificial hybrid between two species of grasses, the blue wild rye (*Elymus glaucus*) and the squirrel tail grass (*Sitanion jubatum*). Furthermore, the origin of a natural species of larkspur (*Delphinium gypsophilum*) occurring in central California is most easily explained on this basis (Figure 6-4). Since the few experiments which have been performed to test this hypothesis have yielded positive results and since segregational sterility based upon many genetic differences is a widespread barrier between species of higher plants, it is possible that a large number of plant species have originated in this fashion.

THE PHENOMENON OF INTROGRESSION

There are three reasons why F_1 hybrids occurring in nature are more likely to form progeny by back crossing to one of the parental species than by mating with each other. First, the hybrids are always fewer, and usually much fewer than are individuals of the parental species. Secondly, the high fertility of parental individuals as compared to the low fertility of hybrids will make an enormously greater number of parental gametes available for fertilizing the hybrids. Finally, back cross progeny, since they will contain genes derived principally from one of the parental species, are the most likely to be well adapted to a habitat already present, which in most instances will be a modification of one of the parental habitats.

For this reason, the commonest result of hybridization is back crossing to one of the parental species, followed by selection of genotypes having a preponderance of genes derived from one species, but a few chromosomal segments introduced from the foreign species. This sequence of events—

A
B
C
A
B
C
D. gypsophilum
D. recurvatum
D. hesperium pallescens

hybridization, back crossing, and stabilization of back cross types by selection—is known as INTROGRESSION. In actively evolving plant species, it is one of the commonest sources of new variability.

The best documented example of introgression is found in a cultivated species, corn. In Mexico, a weedy grass known as teosinte (*Zea mexicana*) is often found in or near fields of cultivated corn or maize (*Zea mays*). The two species are easily hybridized, and the hybrid is partly fertile. Recent archeological discoveries have shown that the ancestor of corn was a wild grass which had most of the characteristics of modern corn, but which was much smaller and had tiny ears. In a succession of layers which have been dated by the radio carbon method, corn cobs have been found which look progressively more like the modern plant. In addition, these cobs contain an increasing number of characteristics in which they resemble teosinte; a resemblance which has been known for some time to be characteristic of some modern varieties of corn. Plants resembling them have been produced artificially by crossing with teosinte the most primitive varieties of modern corn available and making suitable selections from such hybrids. By this means, the importance of introgression with teosinte as a source of some of the more valuable genes found in many modern cultivated varieties of corn has been clearly demonstrated.

Although it does not include evidence from fossils, the circumstantial evidence that introgression has played an important role in the evolution of many wild species of plants is almost as strong as in the corn-teosinte example. In the annual sunflowers, for example, the origin of new races or subspecies through introgression with other species has been repeated by artificial crossing and back crossing. Fewer examples are known in animals, but one exists in domestic cattle. In the high mountains of central Asia occurs a relative of cattle, the yak (*Bos grunniens*), which is also domesticated. Many of the herds of cattle found along the western margin of these mountains, in Soviet Central Asia, contain characteristics which show clearly the presence of genes from the yak. This has, apparently, increased their adaptability to the harsh climate of these mountain areas. Similar introgression of characters from the American bison into cattle has met with some success in the herds of western Canada.

THE ORIGIN OF NEW CHARACTERISTICS VIA HYBRIDIZATION

For the most part, hybridization can give rise only to populations which are intermediate between and share characteristics of the parental populations, and which occupy intermediate habitats. There are, however, three ways in which hybridization can promote the origin of new characteristics.

Fig. 6-4. Drawings showing the leaves, inflorescences, and geographic distribution of three species of larkspur (*Delphinium*) in central California. One of these, *D. gypsophilum* (B), is intermediate between the other two (*D. hesperium* (C), *D. recurvatum* (A)) in morphological characteristics, hybrid fertility, and geographic as well as ecological distribution, and occurs in a habitat which has become available in relatively recent times. Circumstantial evidence strongly indicates, therefore, that *D. gypsophilum* is derived from hybrids between the other two species. From Lewis and Epling, Evolution, 13:511, 1959.

Fig. 6-5. Flowers of *Nicotiana* types: (l. to r.) *N. sanderae; N. langsdorffii;* F_1 hybrid; large-flowered selection, F_8; backcross of large-flowered selection, F_8, to *N. sanderae;* large-flowered selection, F_{10}.

The first is TRANSGRESSIVE SEGREGATION. Since the interaction of genes is not always additive, sometimes a segregating genotype from a wide cross can exceed both of its parents in one or more of its characteristics. A good example is flower size in a hybrid between two species of tobacco. *Nicotiana alata,* a familiar garden ornamental with crimson flowers, has flowers which are among the largest in its genus. On the other hand, the very different looking flowers of *Nicotiana langsdorffii* are among the smallest in the genus (Figure 6-5). The hybrid between the two species is partly fertile, and a great variety of segregating types can be obtained in the F_2 and later generations. After eight generations of selection for large corolla size was practiced on progeny of this hybrid, plants were obtained which had flowers even larger than those of *N. alata,* but which were nearer in color to those of *N. langsdorffii.* Apparently, genes from the small flowered species were able to interact with those of *N. alata* to produce larger flowers than this species ever produces on the basis of its own gene pool.

Another way in which hybridization could increase evolutionary opportunity is through the establishment of NEW GENETIC BACKGROUNDS. Placed on such new backgrounds, mutations which in the parental species were completely inadaptive could now contribute to new adaptive gene combinations. As explained in Chapter 2, one of the principal reasons why mutations occurring in old, well-established species usually reduce the adaptability of their bearers is that nearly all of the mutations which might increase adaptability have already occurred and have been incorporated into the gene pool of the population. A population of hybrid segregates which is entering a new habitat provides, however, an entirely new environment in many ways. In such an environment, many mutations which in the parental species had occurred repeatedly and had been weeded out by selection could take place again and possess an adaptive advantage in combination with other genes already present. Although this is a strong possibility from the theoretical point of view, it is almost impossible to test. The occurrence of new mutations can be recognized only when it takes place on a genetic background which is relatively uniform and homogeneous. In the progeny of genotypes as heterozygous as those produced by hybridization between races or species, there is no way of distinguishing between new mutations and various recombinations of pre-existing genes.

Finally, in some plant hybrids, particularly those between corn and teosinte and between species of cotton, the hybrid condition has INCREASED THE RATE OF MUTATION. Since, as pointed out in Chapter 2, the rate of mutation is not correlated with the rate of evolution, the significance of this increase in mutation rate is hard to estimate. It should, however, help to increase the size of the gene pool from which natural selection, acting in a new environment, could sort out entirely new gene complexes adapted to this environment.

The significance of hybridization in evolution can be summarized by saying that it may provide a rapid increase in the size of the gene pool, from which natural selection for adaptation to a new habitat might quickly sort out entirely new adaptive gene complexes. This potentiality is often realized in plants. In animals, which have much more complex patterns of development, controlled by equally complex and highly integrated sequences of gene action, the proportion of successful recombinants which can segregate in hybrid progeny is much lower. Consequently, the restrictions on successful hybridization are far greater than in plants, and the adaptive recombinations which can segregate from hybrids are far fewer. Because of these facts, most zoologists tend to minimize the importance of hybridization in evolution. One must remember, however, that natural selection, particularly when it is adjusting a population to a new, poorly occupied habitat, can seize on gene combinations which occur at very low frequencies and in a few generations establish them in a population. Hence there are reasons for believing that hybridization has played a larger role in animal evolution than many zoologists think it has. This may be particularly true of hybridization between subspecies and closely related species, in which hybrid inviability and sterility may be only weakly developed.

THE ROLE OF POLYPLOIDY IN PLANT EVOLUTION

In higher plants, adaptive hybrid combinations are often stabilized by polyploidy. This process consists of doubling the chromosome number so that every chromosome present in the original diploid is exactly duplicated in the polyploid hybrid. If, therefore, the hybrid is sterile only because its parental chromosomes are too dissimilar to pair properly, this difficulty is removed in the hybrid polyploid or allopolyploid, and the plant is therefore fertile. Furthermore, since pairing in such plants is entirely or chiefly between two chromosomes derived from the same parent, segregation of the parental characteristics is eliminated, and the polyploid breeds true for its intermediate condition.

One of the best known examples is the rado-cabbage, derived from the sterile hybrids between the cultivated radish (*Raphanus sativus*) and the cultivated cabbage (*Brassica oleracea*). These two species differ radically in

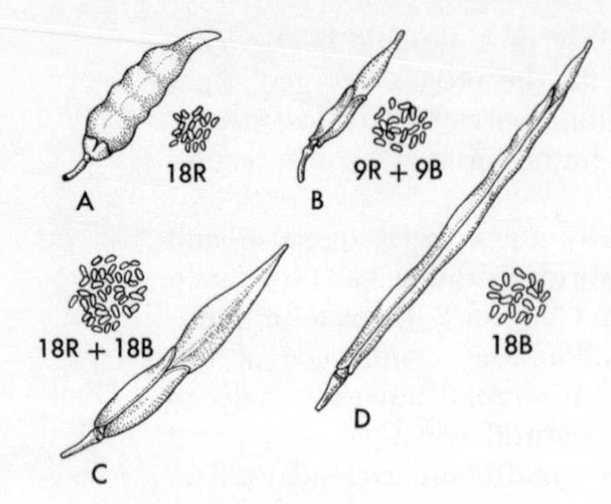

Fig. 6-6. Seed pods and somatic chromosome complements of radish, cabbage, the diploid hybrid between them, and the stabilized, true breeding new species produced by doubling the chromosome number of the sterile hybrid. From Karpechenko.

their fruits, (Figure 6-6) and their chromosomes are so different that they cannot pair at all in the undoubled F_1 hybrid. Consequently, although both parents from nine pairs of chromosomes at meiosis, the hybrid contains 18 unpaired chromosomes at this stage, which are distributed irregularly to the gametes. Doubling the chromosome number produces 18 pairs of similar chromosomes, so that meiosis in the allopolyploid is regular. The plant is fertile, and breeds true for the peculiar, intermediate shape of its pod.

In other examples, such as the hybrid between two species of primroses, *P. verticillata* and *P. floribunda*, the parental species are closely enough related to each other so that the chromosomes can pair in the undoubled F_1 hybrid. These chromosomes are, however, sufficiently different in structure so that crossing over and chromosomal segregation at meiosis produce gametes with disharmonious gene combinations resulting in segregational sterility. When the chromosome number is doubled, the exactly similar chromosomes derived from the same parental species pair preferentially with each other so that the doubled hybrid is fertile and does not segregate in the direction of either parental species (Figure 6-7).

Among natural, wild polyploid species, there are many examples which resemble the rado-cabbage, others which are like the primrose example, and still others which are intermediate between the two in various ways. In addition, some natural polyploids are derived by doubling the chromosome number of a single species, usually a cross between two of its subspecies. These polyploids have four sets of similar chromosomes. This makes their segregation much more complex, and in progeny of a cross between two different autopolyploids of this type, intermediate genotypes are much more frequent relative to the parental genotypes than they are in a cross between two diploids. In this way, any hybrid vigor which may exist in interracial hybrids is buffered by the effects of polyploidy and is more easily maintained by natural selection than it is in diploid hybrids.

Once polyploidy has begun to develop in a genus of plants, it is likely to be a continuing process. A doubled hybrid derived from two original diploid species is called a tetraploid. It may cross back to either one of its parents or to a third species to form a sterile triploid. If this hybrid is doubled, a fertile hexaploid results. Two different tetraploids may similarly cross. Doubling their hybrid will produce an octoploid. In some groups of plants, this trend has continued even further, so that some species have chromosome numbers as high as ten times that of their original progenitor.

Polyploidy is a very common method of evolution in higher plants. Between one-fourth and one-third of the species of flowering plants are polyploid with reference to their nearest relatives. Familiar examples among crop plants are wheat, oats, potato, tobacco, cotton, alfalfa, and most species of pasture grasses. Familiar weeds and wild flowers which are polyploid are the eastern blue flag (*Iris versicolor*), meadow rue (*Thalictrum* spp.), some species of wild rose, chickweed (*Stellaria media*), miner's lettuce (*Montia perfoliata*), yarrow (*Achillea* spp.), and various species of violets and asters. Furthermore, there is evidence that polyploidy in the remote past has given rise to many genera and groups of genera such as the apples, olives, willows, poplars, and many genera of ferns.

Nevertheless, polyploidy has contributed little to progressive evolution. In genera which contain both diploids and polyploids, the major trends of evolution are all represented by diploid species, and the polyploids serve merely to multiply the variations on certain particular adaptive "themes." This is probably because the large amount of gene duplication dilutes the effects of new mutations and gene combinations to such an extent that polyploids have great difficulty evolving truly new adaptive gene complexes.

A

C

D

Fig. 6-7. Plants and meiotic chromosomes of two species of primrose, *P. verticillata* and *P. floribunda*, and the tetraploid species (*Primula kewensis*) produced by doubling the chromosome number of the hybrid between them. Photograph of plants supplied by the late A. B. Stout, New York Botanical Garden; chromosome drawings from M. B. Upcott in Journal of Genetics.

Polyploidy is rare in animals, and is confined almost entirely to forms which reproduce asexually. This is partly because the polyploid condition complicates the segregation of the sex chromosomes, so that many partly or wholly sterile intersexes are produced. In addition, many polyploid animals have disturbed developmental processes, or internal organs (kidneys, nerve centers), which function more poorly than in the related diploids. These forms, even when produced artificially, are hard to keep alive.

Polyploidy is a very striking phenomenon in higher plants, and studies of polyploids can provide much information about the past history of plant groups. This is because it is one of the very few evolutionary trends of which the direction can be clearly determined without fossil evidence. Furthermore, since polyploids can quickly be produced in a single step, many stages of polyploid trends can be reproduced under controlled experimental conditions. Nevertheless, the strong limitations on polyploid evolution render it of only secondary importance in considerations of the phenomenon of evolution as a whole.

Chapter Summary

An evolutionary definition of hybridization is crossing between members of populations having different adaptive gene complexes. When such hybridization takes place, the majority of the progeny in later generations are more poorly adapted to any environment than were the parents of the hybrid. Nevertheless, a small proportion of segregates from such hybrids may be better adapted to a new environment than were any members of the parental population. Consequently, when hybridization takes place in well filled, undisturbed habitats, the hybrids are quickly eliminated and no important results occur. If, however, hybridization is accompanied by the opening up of new habitats, hybrid derivatives may enter these habitats and establish new adaptive gene combinations which contain genes derived from both of the parental species. In modern biota, examples are found most often in habitats disturbed by man, or which were strongly affected by the advance and retreat of glaciers during the ice age.

For hybrid progeny to be successful, they must become stabilized either through establishment of true breeding intermediate gene combinations, backcrossing and selection, or, in plants, doubling the chromosome number. When reproductive isolating mechanisms are determined by many genes, the possibility exists that a new gene combination will be formed which will form partly sterile hybrids when back crossed to either of the parental species populations and thus can become the beginning of a new species. This has been the origin of some species of plants.

New characteristics can arise as a result of hybridization through 1. transgressive segregation, 2. the establishment on new genetic backgrounds of mutations which were consistently rejected by selection on the backgrounds of the parental species, and 3. the effect of hybrid gene combinations which stimulate the occurrence of new mutations.

In higher plants, many new species and some genera have originated by means of hybridization followed by chromosome doubling. This is called hybrid polyploidy. It produces many species which are variations on an established theme of adaptation, but rarely gives rise to new modes of adaptation.

Questions for Thought and Discussion

1. In what way or ways will the environment in which the process takes place affect the outcome of interspecific hybridization?

2. In what ways can the progeny of interspecific hybrids become stabilized so that they will breed true and give rise to new adaptive populations?

3. Discuss the ways in which new characteristics can arise in the progeny of hybrids.

Major trends of evolution

CHAPTER 7

In the preceding chapters, we have considered processes of evolution at the levels of population, race or subspecies, and species. Most people, however, think of evolution in terms of the origin and development of major groups of organisms such as flowering plants and vertebrates, and of continuously evolving lines such as horses and the ancestral lineage of man. The principal objective of the present chapter is to project to this higher level the knowledge and theories which have been developed at the lower levels already considered.

THE CONCEPT OF EVOLUTIONARY TIME

The principal new dimension which enters at this level is that of evolutionary time. During historical times, a few subspecies have evolved in animals such as the house mouse and domestic rabbit, while some kinds of species of plants have evolved in nature and have been synthesized in the laboratory and garden during the present century. Hence at the levels of subspecies and species both the processes and the course of evolution can be followed by human observers. For the origin of all higher categories, however, much more time is required. So far as is known, all modern genera are at least a million years old, and most of them are very much older. A minimum age for modern families is twelve to fifteen million years and for orders about fifty million years. These are figures for mammals, the group which has been evolving the most rapidly during more recent times. For more slowly evolving groups such as insects, molluscs, and flowering plants, these figures should be many times as great. Obviously, we cannot understand the events which have taken place over these enormous extents of time without knowing something about the organisms which existed in past ages. The present chapter must, therefore, consist of a synthesis of facts gathered from the fossil record added to those which have already been presented about modern groups of organisms.

In discussions of past fossil sequences, we are accustomed to mentioning casually the names of past geological epochs and periods, and we also talk about ages of fossils in millions of years. Even scientists who do this, however, are often unaware of the tremendous difference between, on the one hand, the time scale of historical dates and lengths of generations of organisms, and on the other, the scale which must be adopted by those studying the evolution of the major kinds of animals and plants. The following fantasy may serve to emphasize this point. Suppose that all of the events of modern history, from the birth of Christ up to the present, were compressed into a television show lasting for an hour. And, suppose that the earlier events of human prehistory and the course of organic evolution were presented in a series of one hour shows at the same time scale, i.e. each show representing two thousand years. Suppose, furthermore, that these shows were presented during every waking hour; 16 per day, and every day of the week. How long would it take to cover the various events of evolutionary time?

The prehistory and evolution of man since he acquired his present bodily form (about 100,000 years) would require about 3 days. The evolution of the lineage of man since it diverged from that of the apes (perhaps about 10,000,000 years) would require ten months (313 days) to present. The course of evolution of such families as those of the cat, dog, and horse (35 to 55 million years) would require $3\frac{1}{2}$ to $5\frac{1}{2}$ years. The entire continuous fossil record (600,000,000 years) would require about 60 years, and the entire course of the evolution of life between 250 and 400 years. Paleontologists speak of evolutionary changes which occur during a period of one to two million years as "rapid." In terms of the events of human history, changes occurring at this rate are extremely slow. They are even slower in terms of the lengths of generations of animals and plants, on the basis of which the processes of evolution up to now have been considered. Another comparison, in which the example chosen is the evolution of the size of the human brain, is presented in the next chapter.

THE FOSSIL RECORD

The record of past forms of life is now extensive and is constantly increasing in richness as paleontologists find, describe, and compare new fossils. Good descriptions of evolutionary history as revealed by the fossil evidence can be found in a number of books. A penetrating analysis of rates and processes of evolution on the basis of this evidence has been made by Dr. G. G. Simpson in his book, *The Major Features of Evolution.* In the present discussion, only those features will be mentioned which pertain directly to the problem of relating the processes of evolution to the origin and development of major groups of organisms.

In the best-known groups of organisms, the fossil record as we now know it tells us in a general way the time when the major groups of organisms appeared. In some instances, it also tells us what the forerunners of modern orders and families were like. Less often, fossil sequences can be found which trace the successive appearance of families within an order and of genera within a family. But, when they try to relate such events to the processes which have given rise to races and species, evolutionists are impressed above all with the imperfection of the fossil record for this purpose.

This fact cannot be overemphasized. Although many thousands of different kinds of fossils have been discovered, they are still only a tiny fraction of the organisms which have existed. Dr. Simpson estimates that the available fossils represent only a fraction of one per cent of the species which have existed during the evolution of life. Not only is the record very incomplete, but it is, in addition, a strongly biased sample. In most places where organisms die, their remains are quickly destroyed by other organisms, particularly the bacteria and fungi which cause decay. Since fossils are preserved chiefly under water, or in water soaked ground, nearly all of the deposits of terrestrial animals and plants are in or near ancient river and lake beds. This means that organisms which normally live in such habitats have a much better chance of being preserved than have those which live in uplands or on mountain sides. Moreover, periods of the earth's history when shallow seas, lakes, and large rivers were widespread are more strongly represented in the fossil record than are those periods when mountain ranges were most highly developed.

Another source of imperfection is that almost no organisms are preserved in their entirety. Soft parts are very rarely preserved in fossils. The larger and harder is the animal or certain of its structures the greater is the chance of its being preserved. For following the evolution of vertebrates, this bias is less of a handicap than it is for other forms of life. The parts of these animals which are most often preserved are their teeth and bones. These parts enable paleontologists to make many deductions about the way in which the animals lived. Their teeth are modified according to their diet while their method of locomotion is revealed by their foot structure. Moreover, their nervous reactions are revealed by casts of their skulls and positions of the major sense organs, and their intelligence is related, though very indirectly, to the size of their brains. These structures are also preserved in many vertebrate fossils. Nevertheless, the record shows that most of the animals which were ancestral to the major orders and families were relatively small. The larger animals, which were preserved most often, represent highly evolved members of their particular lines and tell us little about how one major adaptive complex of characters evolved from another. In other groups of organisms, the accidents of preservation have been much less favorable. Insects, since they are not only small but also very fragile for the most part, are preserved only in a few deposits, widely separated from each other in space and time. In many of these the animals are poorly preserved. Plants are most often represented by leaves, wood, and particularly pollen, parts which botanists have found in modern plants to be much less useful for classification and for following trends of evolution than are the larger reproductive parts.

An additional weakness of the fossil record is that many past epochs are represented by fossil beds which were being deposited simultaneously in only a small number of different regions of the earth. For understanding the origin of modern species the facts of geographic distribution, such as allopatry and sympatry of related populations, are of the greatest importance. Such facts can rarely be obtained from studying the fossil record.

The bias inherent in the fossil record is exactly of the wrong kind for evolutionists who wish to learn how the major groups of organisms originated. In modern floras and faunas the greatest diversity of races and species exists in mountainous habitats, where climate, soil, and other factors can vary greatly in a small area. In such regions, also, new habitats are opened up much more often than in flat lowlands. Consequently, stages in the origin of modern species can be found most easily in upland regions, or on islands which are hilly or mountainous. These are, however, the places in which the chances that fossils will be preserved are the lowest. The facts which we know about processes of evolution in modern times, combined with our knowledge about the ways in which fossils are preserved, would lead us to expect that groups of animals and plants which have reached the climax of their evolution and are exploiting to the full highly evolved and well established adaptive complexes will form the great bulk of fossils. Those organisms which are experimenting with new ways of life and show transitions between the major groups are much less likely to have been preserved.

In view of these imperfections, the fact is surprising that paleontologists have been able to supply a great deal of information regarding the course of evolution. Moreover, this information is most abundant in respect to the vertebrates, which include our own ancestral line. This is why the vertebrate record will be used almost exclusively in the rest of the present chapter.

THE DATING OF FOSSILS

In order to determine the relative age of fossils, paleontologists rely first on regions in which fossils are most abundant. As stated earlier, these are likely to be regions which have been lowland areas of deposition, and have been little disturbed by mountain building movements. The older part of the record, the Paleozoic Era (see Table 7-1), is very well represented in parts of the central and eastern United States (Illinois, Indiana, Ohio, Kentucky, Tennessee, West Virginia, Pennsylvania, and New York). There, deposits which accumulated during a period of 250 million years are found in an almost regular succession. The last 100 million years of the record, extending from the middle of the Cretaceous Period into modern times, are well represented in the western part of our country. These deposits were, however, explored much more recently than were those of Europe. Consequently, when the fossil record was first studied, names were given to the deposits found in various parts of Europe, and the relative ages of European fossil beds were first determined. These names have now become standardized and are used to designate the relative age of deposits all over the world. Table 7-1 lists the various eras, periods, and epochs which constitute the universally used system.

When successions of fossil beds in the better preserved layers of different continents are compared, they are found to contain certain genera in common. Particularly in the oceans, the faunas existing at any one time in different parts of the earth were much alike and differed in the same way from those which preceded and followed them. This is why the names based upon European deposits can be successfully applied to deposits found all over the earth. In mountainous regions, fossil bearing strata may in some places be

TABLE 7-1. The Geological Periods and Epochs.

ERA PERIOD-EPOCH	MILLIONS OF YEARS AGO WHEN IT BEGAN	PRINCIPAL EVENTS WHICH OCCURRED
CENOZOIC		
Quaternary-Pleistocene	1	Ice age, evolution of man.
Tertiary-Pliocene	11	
Miocene	25	Spread of anthropoid apes.
Oligocene	36	Origin of more modern families of mammals.
Eocene	54	Origin of many modern families of mammals.
Paleocene	65	Origin of most modern orders of mammals.
MESOZOIC		
Cretaceous	135	Appearance of flowering plants; extinction of dinosaurs at end; appearance of a few modern orders and families of mammals.
Jurassic	181	Appearance of some modern genera of conifers; origin of mammals and birds; height of dinosaur evolution.
Triassic	220	Dominance of mammal-like reptiles.
PALEOZOIC		
Permian	280	Appearance of modern insect orders.
Pennsylvanian	310	Dominance of amphibians and of primitive tropical forests which formed coal; earliest reptiles.
Mississippian	355	Earliest amphibians.
Devonian	405	Earliest seed plants; rise of bony fishes.
Silurian	425	Earliest land plants.
Ordovician	500	Earliest known vertebrates.
Cambrian	600	Appearance of most phyla of invertebrates.

Before this time, very imperfect fossil record, and time of appearance of different forms of life is uncertain.

folded, broken, and thrust over each other so that older fossil beds may occasionally lie on top of younger ones. Although such deposits are often difficult to date, their age in relation to neighboring deposits can usually be determined on the basis of certain key fossils which they contain.

We must always remember, however, that the dating of fossil beds is always approximate, and becomes less precise the older are the beds. In dating the older beds, which contain the forms most critical for determining how major groups of organisms arose, errors of one to two million years are to be expected. Consequently, the fossil record of these events will never be sufficiently complete or precisely understood so that we can interpret them in terms of the evolutionary processes which we are following in modern biota, except in a general, indirect fashion.

The principal ways by which scientists determine the absolute age of fossil deposits in millions of years are all based upon the fact that radioactive elements decay at regular rates, and so form "radioactive clocks." The particular "clock" which is most useful depends upon the age of the deposit which is being dated. For very recent fossils, 40,000 years old or less, the decay of an isotope of carbon known as carbon-14 is by far the most useful. All living beings contain a constant amount of this isotope, which starts to decay when the organism dies and no longer has its carbon replenished from outside. For the great bulk of fossils, however, a more useful method relies on the decay of uranium-238 to form lead-206. This method can be used only for igneous rocks, which were formed by the cooling of molten masses of minerals far below the surface of the earth. When each rock mass cooled and crystallized to assume its present mineral composition, those minerals containing U-238 formed separate crystals which contained no lead. Since the rock was formed, the uranium has been decaying to form lead at a regular rate, so that the higher is the proportion of lead in its uranium bearing minerals, the older is the rock.

There are several difficulties with this method. One is that uranium-containing rocks are relatively uncommon. In addition, igneous rocks, by their very nature, rarely contain fossils. The ages of fossil-bearing sedimentary rocks can be dated according to the uranium-lead "clock" only by indirect comparisons with neighboring igneous rocks. More recently, "radioactive clocks" based upon isotopes of potassium and argon have increased the number of reference points for dating fossil beds, and we can expect that still more such "clocks" will be developed in the future. An encouraging feature of these methods is that they have supported the estimates of the relative lengths of different geological periods which paleontologists had previously made on the basis of more indirect methods. At present, the dates given in Table 7-1 are generally accepted by geologists and paleontologists as accurate within a few millions of years, at least in respect to the more recent dates.

THE ORIGIN OF HIGHER CATEGORIES

Before answering the question: "Did the higher categories evolve by means of the same processes which gave rise to races and species?", we must first be clear in our minds as to what we mean by "higher categories." According to the system devised by Carl von Linne two hundred years ago and now in universal use, organisms are arranged into a hierarchy of categories, of which the lowest is the species and the highest the kingdom. Between them are placed the genus, family, order, class, and phylum. These categories are sometimes subdivided still further, but only the major ones mentioned above will be considered in the present discussion. Table 7-2 presents the classification of the dog according to this system. We know enough about dogs and their relatives to say that the processes of mutation, genetic recombination, natural selection, and reproductive isolation, as discussed in previous chapters of this book, can explain the differentiation from a common ancestor of the various species of the genus *Canis*, such as the dog, wolf, coyote, and jackal. We must now ask the question: Can these processes, acting over the millions of years encompassed in the evolutionary time scale, also account for the differentiation of dogs from foxes, of dog-like from bear-like animals, of carnivorous from herbivorous mammals, of mammals from reptiles, of vertebrates from other kinds of animals, and of animals from plants and microorganisms?

Our answer to this question depends in part on how we answer another, somewhat philosophical question: Are the categories of the systematic hierarchy intrinsic entities which the naturalist merely discovers, or are they groups which naturalists themselves have established in order to understand better the complex pattern of living beings in nature?

The correct answer to this question, which has now become reasonably clear, can be understood on the basis of the following predictions. If genera, families, and other categories are intrinsic entities which only need to be discovered, then we would predict that the more intensively a group was studied by different biologists, the more easily could these authorities come to agreement on the limits of the categories. On the other hand, if higher categories are largely human inventions, then each biologist approaching the problem of classifying a particular group of organisms would have a somewhat different idea from his predecessors and contemporaries as to which characteristics are the most important and would define the categories in a somewhat different way. Consequently, as more biologists acquired more facts about a group of organisms, they would not be able to define its cate-

TABLE 7-2. Classification of the Dog According to the Commonly Used Categories.

CATEGORY	NAME	RELATED CATEGORIES OF THE SAME RANK
Species	*Canis familiaris* (L.)	*Canis latrans;* coyote.
Genus	*Canis*	*Vulpes;* fox.
Family	Canidae	Ursidae; bears.
Order	Carnivora	Rodentia; squirrels, rats, porcupine.
Class	Mammalia	Reptilia, Aves (birds), Amphibia.
Subphylum	Vertebrata	Tunicata; sea-squirt.
Phylum	Chordata	Arthropoda; crustaceans, insects, spiders.
Kingdom	Animal	Plant

gories any better, and the differences of opinion between them might even become stronger. If, therefore, greater familiarity with a group does not make its subdivisions easier to define, then we must conclude that these categories are largely human constructs, created by naturalists in order to make classification easier.

When we look at different groups of organisms, we find that for some of them the first prediction has been realized, and for others, the second prediction has come true. The orders of mammals, such as bats (Chiroptera), primates, rodents, whales (Cetacea), carnivores, etc., have been recognized as such for more than a hundred years, and modern knowledge has increasingly confirmed, with few exceptions, the classical delimitations of contemporary orders. Within many of the orders, such as the whales, carnivores, elephants, odd-toed and even-toed ungulates, the modern families are equally well defined. In other orders, particularly the rodents, zoologists have had much greater difficulty in deciding how its families can most naturally be delimited. In respect to genera, differences of opinion extend to many families, and universal agreement is confined largely to those families in which the modern genera are relatively few and small.

These differences in the ease of defining categories can be understood when we look at the fossil record. On the whole, the mammals are a declining class. They reached their peak in the Miocene and Pliocene epochs. Since then, the number of old genera to become extinct has exceeded that of new genera which have evolved in most orders. Conspicuous exceptions are the rodents and some groups of even-toed ungulates. It is exactly in these exceptional groups which are still flourishing that the delimitation of genera is the most difficult. This fact would suggest that higher categories become well defined through the extinction of populations or species which are intermediate between the most successful surviving groups.

When we review the classification of other groups of organisms, we find the same relationship between distinctness of modern categories and probable extinction of intermediate groups. In higher plants, genera are most easily defined in the pine family and other cone bearing trees, as well as in the magnolias and their relatives. These groups are known to be relatively ancient. Some of their genera are clearly declining or are small persistent relics of formerly more widespread groups. On the other hand, in plant families which are known to have increased in importance during the more recent geological epochs, and which have spread still further as a result of man's activity, genera are particularly hard to define. The best examples are the grass family and the sunflower family. Apparently, therefore, the process which increases in importance as we consider the evolution of higher categories is extinction. This process results inevitably from an extension of evolution into time spans comprising millions of years.

If categories become well defined because forms intermediate between them become extinct, then in the history of groups having a good fossil record we should be able to find periods when categories which are now well defined were connected by transitional forms. If we analyze the fossil record of vertebrates, this is exactly what we see. Among modern animals, the dog and bear families are regarded as definitely related to each other, but even when all contemporary members of the two families are considered, nobody has any difficulty in distinguishing bears from dogs, foxes, and coyotes. In the Miocene and early Pliocene epochs, however, the situation was different. At that time, animals intermediate between dogs and bears were common, so that paleontologists have great difficulty in deciding just when the dog and bear families became distinct from each other (Figure 7-1). The present distinctness of the two families is due partly to the fact that bears, in connection with their acquisition of an omnivorous diet rather than one consisting only of meat, have acquired distinctive jaws, teeth, and faces. At the same time, their relatively large size and slow movements are associated with the fact that they no longer pursue their prey, as did their ancestors. This adaptive radiation, however, would not have made them easy to recognize as a separate family unless the intermediate dog-bear animals which existed in the Miocene and early Pliocene epochs had become extinct since then.

Going farther back in the fossil record, we learn that in the latter part of the Eocene epoch, primitive animals which are now clearly recognized as forerunners of the principal families of carnivores; dogs, cats, weasels, civets, and their relatives, were linked together by a complex network of resemblances. When the characteristics by which we now recognize these major families first appeared, they were distinctive of genera rather than families. As in the example of dogs and bears, the distinctness of the other major modern families of carnivores was a result of the combined effects of continued adaptive radiation by means of genetic change guided by natural selection plus extinction of generalized intermediate forms. The same trends can be followed in the evolution of families in other orders of mammals and, though less clearly because of imperfection of the record, the differentiation of the orders themselves.

Although the differentiation of the classes of vertebrates is less clearly illustrated by the fossil record, the available evidence indicates that their origin was no different from that of orders and families. In the Devonian period, one of the most abundant and dominant groups of fishes was the

Fig. 7-1. Skulls of (A) a modern wolf, (B) a bear, and (C) an intermediate form which lived during the Miocene Epoch.

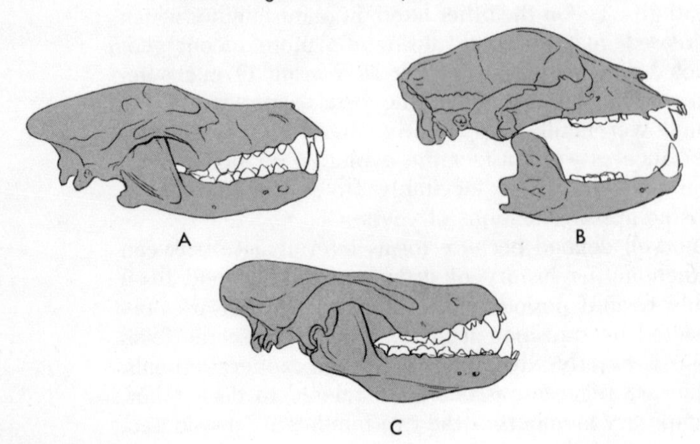

Fig. 7-2. An extinct lobe fin fish (Crossopterygian), which lived during the Devonian Period, together with a primitive amphibian which was derived from it or a similar form. From Romer, *The Vertebrate Story* (University of Chicago Press).

lobe-fins. These fishes possessed two pairs of fins which in their center contained thick, fleshy lobes, of which the skeletal bones were not unlike those of primitive amphibians (Figure 7-2). We cannot reconstruct the way in which these ancient, extinct fishes lived, but since some modern fishes use their fins for waddling slowly over the bottom of shallow bodies of water, we can suspect that the lobe fins did the same. They also had openings for nostrils in the roof of their mouth. This fact, along with the close relationship of lobe-fins to modern lungfishes, suggests that the lobe-fins had lungs, and breathed air when their water became foul, stagnant, and lacking in oxygen.

At the end of the Devonian Period appeared the first animals with legs rather than fins, the most primitive amphibia. These animals were, however, by no means land dwellers. All of the adaptations of their skeletons, aside from their limbs, were for life in the water. Although they probably made short journeys overland from one body of water to another, their limbs were probably used more for waddling over the bottoms of swamps and shallow pools in search of aquatic prey than for life on land. Thus, the structures which eventually made life on land possible, lungs and feet, were probably first acquired by basically aquatic animals as part of an adaptive radiation into the extensive bodies of shallow water which existed when these animals first evolved.

Although later amphibians, like some of our modern frogs, toads, and salamanders, evolved adaptations which enabled them to live all of their adult life on land, all amphibians require water or at least moisture for their reproductive stages. This is because their soft, jelly-like eggs have no protection against drying up. The first truly terrestrial class of vertebrates is the reptiles, which lay eggs protected by shells and containing large amounts of food material for nourishing the early embryo. Since eggs are only very exceptionally preserved as fossils, we have no way of finding out how the reptilian egg evolved. There are, however, many differences between modern reptiles and amphibia in the structure of their skeletons, and these have been used by paleontologists for recognizing the first reptiles to appear. An eminent paleontologist, A. S. Romer, remarks of these animals: "Primitive Paleozoic reptiles and some of the earliest amphibians were so similar in their skeletons that it is almost impossible to tell when we have crossed the boundary between the two classes." (*Vertebrate Paleontology,* p. 121). In all likelihood a zoologist transported to the early part of the Permian period, not knowing anything about the animals which evolved later, would have placed in the same class, the same order, and perhaps even in the same family animals which had some characteristics of amphibians along with others having characteristics now associated with reptiles.

In respect to the early evolution of mammals, the same situation exists. The distinctive characteristics of modern mammals; warm blood, hair, and the ability to suckle their young, cannot be determined in fossils. In respect to their skeletons, however, modern reptiles are, and the dinosaurs were, very different from modern mammals. On the other hand, the animals which dominated the land in the later Permian and early Triassic Periods, before the dinosaurs appeared, were the mammal-like reptiles or therapsids, which in both their skulls and teeth were almost halfway between typical reptiles and primitive mammals (Figure 7-3). Although their soft parts are completely unknown, some features of the skeletons of therapsids suggest that at least the most advanced members of the group were not completely cold blooded, but had some form of temperature regulation. Since the most important function of hair is protection of a warm blooded body, we may speculate also that the evolution of hair took place within the therapsids. Moreover, the most primitive modern mammals, the duckbill and the spiny ant eater of Australia, have both skeletal characteristics and chromosomes which set them off sharply from other mammals. Although their fossil ancestry is completely unknown, most paleontologists suspect that they have descended from the therapsids independently of other groups of mammals. Since these Australian monotremes possess hair, warm blood, and suckle their young in a primitive fashion, the speculation that these animals were independently descended from therapsids carries with it the implication that some members of this group of "reptiles" resembled primitive mammals more than modern reptiles in the soft parts of their bodies. If the therapsids could be resurrected, we might again find a group in which characteristics now diagnostic of different classes were separating forms which were genetically related to such a degree that they would naturally be placed in the same order or even family.

During the Triassic period, the therapsids gave rise to a group of rather small, light boned, and apparently very active animals, the ictidosaurs. These animals, which existed for at least twenty million years during the later Triassic and early Jurassic periods, had skeletons which were mammal-like in every respect except for two small bones of their lower jaws, described later in the chapter (Figure 7-9), and which in mammals have become two of the small bones in the middle ear. Since the paleontologists who classify the skeletons of vertebrates have arbitrarily decided that the presence of these bones on the jaw is a characteristic of reptiles, these ictidosaurs are placed in the reptilian class. Commenting on them, another eminent paleontologist, E. H. Colbert, remarks: "All of which indicates how academic is the question of where the reptiles leave off and the mammals begin." (*Evolution of the Vertebrates*, pp. 134–135).

The first true mammals appeared in the middle Jurassic, about the time when the ictidosaurs were becoming extinct, and were contemporary with the earlier dinosaurs. Consequently the transition from reptiles to mammals is

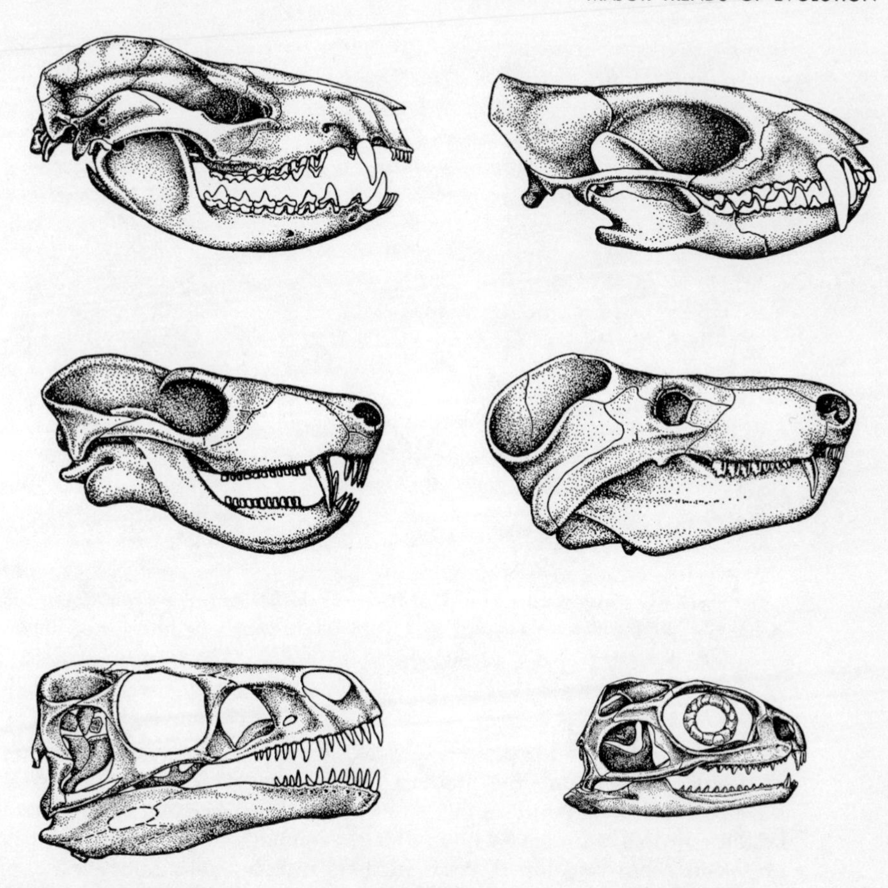

Fig. 7-3. Skulls of mammals, reptiles, and intermediate forms. *Top,* a modern opossum and a primitive placental insectivore (*Deltatheridium*) which lived during the Cretaceous Period. *Middle,* two therapsids, *Bauria* and *Cynognathus,* which flourished during the Triassic Period. *Bottom,* a small early Mesozoid reptile (Ornitholestes), and the modern tuatara (*Sphenodon*), found only in certain islands off the coast of New Zealand. Drawings from Romer, *Vertebrate Paleontology* (University of Chicago Press).

gradual and even, both in time and in the bodily form of the animals themselves.

The transition from reptiles to birds is more poorly documented than are the other transitions between classes of vertebrates. Nevertheless, many of the smaller reptiles in the group ancestral to dinosaurs and crocodiles had light skeletons from which those of birds could have arisen, and moreover walked exclusively on their hind legs, as do birds. Furthermore, the earliest fossil birds, from Jurassic deposits of Germany, had jaws containing teeth and forelimbs with well developed fingers (Figure 7-4). We classify them as birds because feathers are preserved with their skeletons; but if their preservation had been somewhat poorer and the feathers were not present, these animals might well have been classified as reptiles.

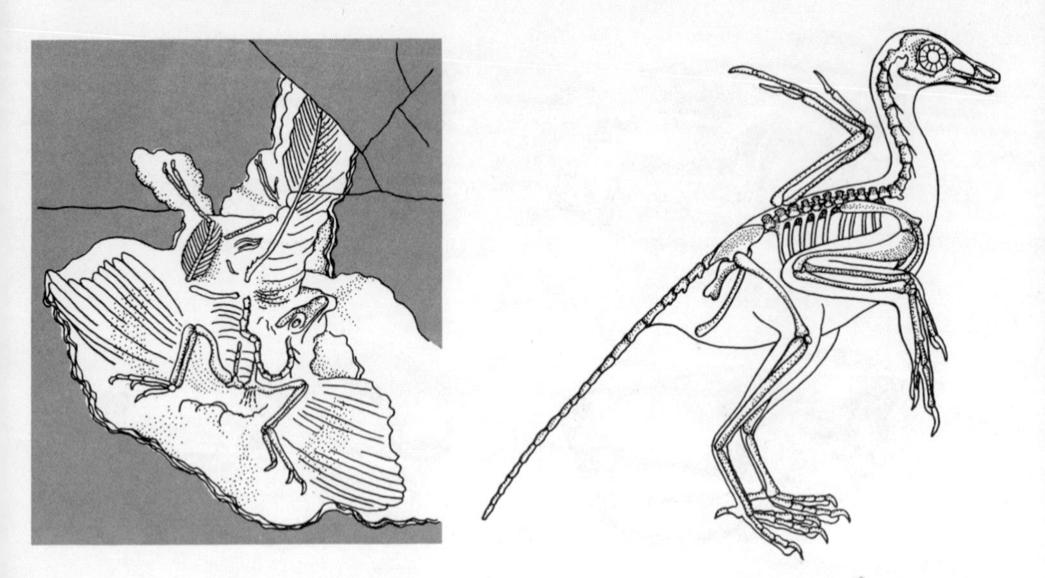

Fig. 7-4. Fossil skeleton of *Archaeopteryx,* a form intermediate between reptiles and birds, which lived during the Jurassic Period. From Romer, *Vertebrate Paleontology* (University of Chicago Press).

Thus, the fossil record of vertebrates strongly suggests that the characteristics which distinguish the modern higher categories appeared first as distinctive features of certain species or genera. They became characteristics of families, orders, and classes only after descendants of the animals which first possessed them developed them further, radiated into numerous adaptive niches, and became separated from other groups by extinction of intermediate forms. In other groups of organisms such as insects and higher plants, in which the fossil record is far more fragmentary, profound gaps exist between many orders, suborders, and classes. Furthermore, no transitional forms are known between any of the major phyla of animals or plants. In view of the incompleteness and biased nature of the fossil record in all of these groups, and the extremely long time, measured in hundreds of millions of years, since the various phyla of organisms evolved, the large gaps which exist between many major categories of organisms aside from the vertebrates are most reasonably ascribed to known imperfections in the fossil record. The hope always remains that new fossil finds will fill in some of these gaps.

To the questions which were asked at the beginning of this section, we can now give the following answers. When the characteristics first appear which later distinguish the major higher categories, they are distinctive of species and genera. At this stage, definition of genera is somewhat arbitrary, and their naturalness is not evident. As members of different groups diverge farther from each other in evolution, and as intermediate organisms become extinct, the higher categories become easier to define. Consequently, the only qualities of naturalness or intrinsic identity which higher categories possess are conferred on them by a continuation of known evolutionary processes through long periods of time, plus the extinction which inevitably accompanies the reaction of communities of organisms to repeated changes of their environment. There is, consequently, no reason to invoke any special processes to account for the evolution of higher categories, beyond those which give rise to races and species.

A further point must be emphasized in connection with the evolution of families, orders, and classes. This is its "mosaic" character. As pointed out in connection with both the evolution of amphibia from fishes and of mammals from reptiles, the various characteristics which now distinguish the more evolved class probably evolved separately, some relatively early, others much later, at periods of evolutionary time which in some instances were separated from each other by millions of years. This is almost certainly true of the basic adaptive characters of the placental mammals: warm blood, hair, the four-chambered heart, the ear apparatus, vivipary rather than egg laying, and the placenta. The placental mammals did not radiate into their present diverse ways of life or become the dominant land animals until all of these characteristics had become perfected. Those characteristics which possess a close functional connection with each other, such as hair, warm blood, and the four-chambered heart, were probably fairly closely synchronized in their evolution. On the other hand, the evidence strongly suggests that the evolution of these physiological characteristics was only weakly correlated with the steps in the evolution of the ear and of the reproductive system.

Consequently, we cannot speak of any single "step" in the evolution of mammals from reptiles. In some instances, such as the change in position of the jaw bones to the ear, a relatively small number of genetic changes may have triggered off the evolution and establishment of a new adaptive complex with respect to that particular character, as in the evolution of mimicking races of butterflies or of species of columbines with spur bearing petals, described in Chapter 4. These changes would, however, have occurred at the level of subspecies or closely related species. A contemporary taxonomist, transported to the Mesozoic era and not knowing anything about the evolutionary future, would probably have classified the first population bearing all three bones; hammer, anvil and stirrup, in its middle ear, as an aberrant species belonging to the then widespread group of therapsid reptiles. As stated above, this group probably already possessed a mixture of characters which we now associate on the one hand with reptiles and on the other with mammals.

In the next chapter, evidence will be presented to suggest that the evolution of man from the apes was probably of this same mosaic nature. Characters evolved in synchronous correlation with each other only to the degree to which they were functionally interdependent.

RATES AND TRENDS OF EVOLUTION

In Chapter 4, three different kinds of natural selection were recognized—stabilizing selection, directional selection, and disruptive selection. These same kinds of selection can also be recognized in the evolution of major groups of organisms. Disruptive selection, followed by simultaneous directional selection in several different lines of evolution, produces the pattern of evolution known as ADAPTIVE RADIATION. This process has already been described in Chapter 5 as being responsible for the differentiation of races within a species. It can, however, take place at any level of the systematic hierarchy. Within many widespread plant species, adaptive radiation of populations into meadow, seashore, cliff, alpine, and other habitats, produces the diversity of races which have been analyzed by many botanists. At the level of species in the dog genus, *Canis*, adaptive radiation, beginning at the level of races and extending to that of species, has produced the timber wolf in the forests of the north, the coyote of our western plains, the jackal in the warm dry

Fig. 7-5. Adaptive radiation of odd-toed herbivorous mammals from the primitive Eohippus (*Hyracotherium*), from the Eocene epoch to the present. From Simpson, *Horses* (Oxford University Press).

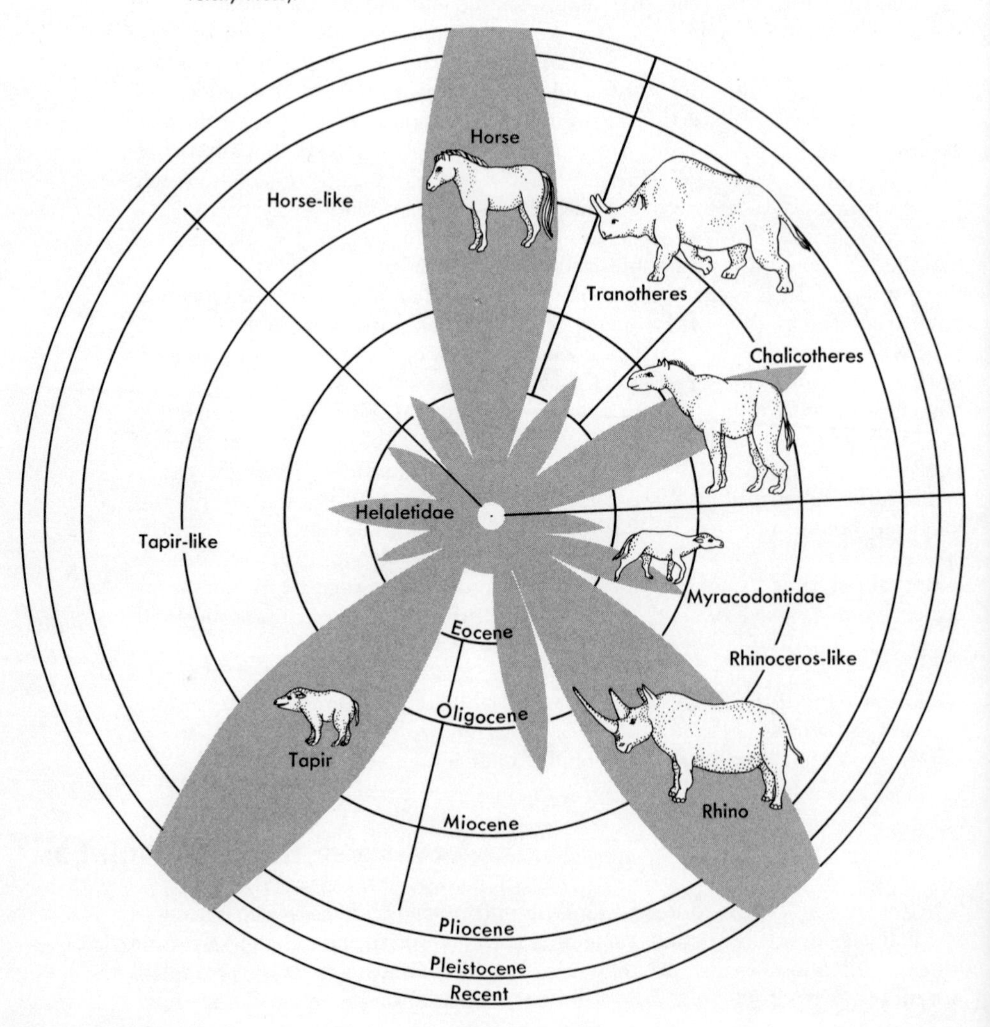

regions of southwestern Asia, and several others. At a higher level, adaptive radiation from generalized mammals which lived at the beginning of the Tertiary Period gave rise to the various modern orders. These were distinguished at first by their food habits (insectivores, carnivores, herbivorous ungulates, etc.) and in the more specialized end products of radiation by various modifications of body form, limb structure, organization of skulls and teeth, and numerous other adaptations to their particular way of life. A typical example of this type of radiation, the evolution of horses, is presented later in the present chapter. Within each order, adaptive radiation at the level of modern families gave rise to many different lines of evolution (Figure 7-5).

Two conditions of the population-environment relationship provide opportunities for adaptive radiation. One takes place when a species enters a new, unoccupied habitat in which diverse ecological niches are available. An example is the evolution of finches (*Geospizidae*) on the Galapagos Islands (Figure 7-6). Probably soon after these volcanic islands rose from the sea and acquired enough vegetation so that birds could live on them, a single pair or a small number of generalized, finch-like birds arrived there from the South American mainland. Since they had no competitors and few if any predators, they must have been able to multiply very rapidly.

During this period of expansion, colonizing "explorers," often consisting of a single pair of birds, must have occasionally made long journeys from one island to another, or from one part to another of the larger islands. In their new home, they would have found a somewhat different environment from any which the species had previously encountered. Furthermore, their descendants could have multiplied while maintaining complete isolation from any related population. In this way, each isolate would develop, by a combination of chance and natural selection (see Chapter 4), a population having a new adaptive complex. After "explorers" from these newly established populations had re-invaded the home of their ancestors, sympatric populations could persist together indefinitely, because they had become adapted to their environment in different ways.

Some of them became adapted to perching on trees and foraging for insect larvae and pupae by removing the bark with their sharp beaks. Others became ground feeders with moderate sized beaks, subsisting on small seeds and slower moving insects, as do modern sparrows. Still others evolved large, powerful beaks, capable of crushing and eating large, heavy coated seeds. One of the most remarkable adaptations is that of the cactus finch, which lives on the large tree cacti that are common in many parts of the islands. These birds pluck long spines from the cacti and, holding them in their beaks, probe the crevices of the cactus stems for the insects which are hidden there, remove them and feed on them. These different types of Galapagos finches are shown in Figure 7-6.

Adaptive radiation will also follow when a population acquires by evolution a new complex of adaptive characters which enables it to exploit an available environment more efficiently than can any of the other organisms already present there. Perhaps the most striking example of adaptive radiation in response to a newly evolved way of life is that of the placental mammals at the beginning of the Tertiary Period. At the end of the Cretaceous Period,

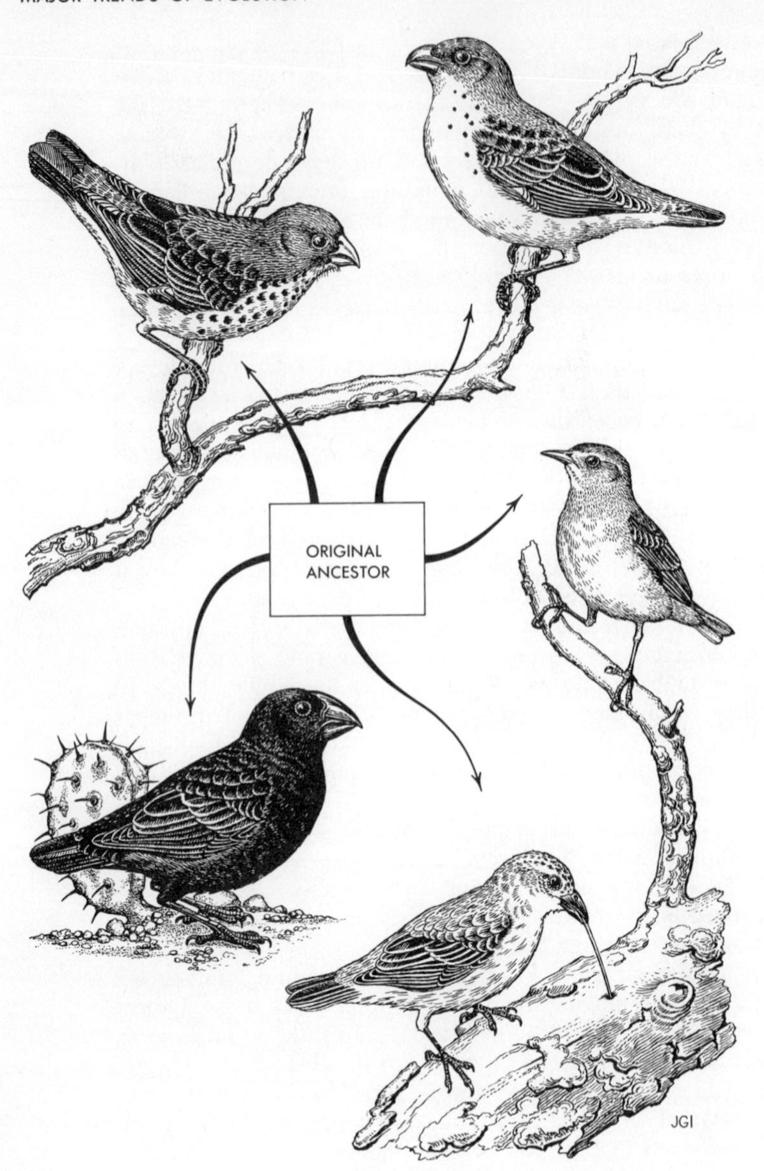

Fig. 7-6. Some of the different species of Galapagos Finches, which originated from a common ancestor through adaptive radiation in association with the different habitats available on the Galapagos Islands. From Lack, *Darwin's Finches* (Oxford University Press).

the dinosaurs, which had dominated the earth for more than a hundred million years, became extinct for reasons which are still not well understood. This left the earth devoid of large land animals, with the exception of tortoises, lizards, and possibly ostrich-like birds. Meanwhile certain mammals, which throughout the age of dinosaurs had been evolving separately and in a subordinate position, had perfected their adaptation of nourishing the embryo and fetus within the uterus of the mother by the evolution of the complex system of tissues known as the placenta. This provides a much more efficient means of nourishing the fetus before birth than exists in the egg-laying reptiles or birds, and probably is also more efficient than the pouch of the marsupials. Furthermore, because the mother can suckle her young from her milk, these babies are given a long period of safe dependence upon a protective mother. This long developmental period is essential for the maturation of the complex brain and nervous system found in mammals, and consequently for their greater mental ability relative to reptiles and most birds.

Equipped with this more efficient reproduction and greater brain power, the placental mammals "exploded" in a series of adaptive radiations which were unprecedented in their rapidity and extent. During the eleven million years which comprised the Paleocene Epoch, the original placentals, which were unspecialized insectivorous animals about the size of a rat, gave rise to hedgehog-like insectivores, gnawing rodents, tree climbing lemurs, soaring bats, primitive carnivores, herbivorous ungulates, and many other forms. Immediately following this initial adaptive radiation, each derived line underwent secondary radiations to produce an even greater diversity of adaptive complexes. During the 22 million years of the Eocene Epoch, these gave rise to the forerunners of many of the modern families of mammals.

There were similar, equally spectacular, adaptive radiations of primitive fishes in the Devonian, primitive reptiles in the Permian, and dinosaur-like reptiles in the Jurassic. Among plants, the most spectacular adaptive radiations were those of the primitive land plants in the late Silurian and early Devonian, and of flowering plants in the Cretaceous Period.

If an evolutionary line has become adapted to an environment which is changing progressively over a long period of evolutionary time, it may progress in accord with this environmental change by means of a series of adaptive radiations. This has apparently been the evolutionary history of the horse family, as revealed by the extensive array of fossils now available (Figure 7-7). It begins with *Hyracotherium* (Eohippus), an animal about eighteen inches long, or the size of a small cat. These "dawn horses" had four toes on their front feet and three toes on their hind feet, on which hoofs were not yet fully developed. Their low crowned teeth, with a relatively simple pattern of enamel, adapted them to browsing on the soft herbage of their forest habitat. They lived in the lower Eocene Epoch, 50 to 54 million years ago.

The series of adaptive radiations which eventually gave rise to the modern horse involved principally the following changes:

1. Increase in size, and particularly in the length and slenderness of the legs, which made for faster running. In harmony with this change, the back became straighter and stiffer.
2. Decrease in number of toes, and expansion of the "toenail" of the single remaining toe, to form a hoof. This helped the later horses to gallop over hard ground.
3. Increase in width of the incisor teeth, height of the molars, along with the development of complex patterns of hard enamel. These

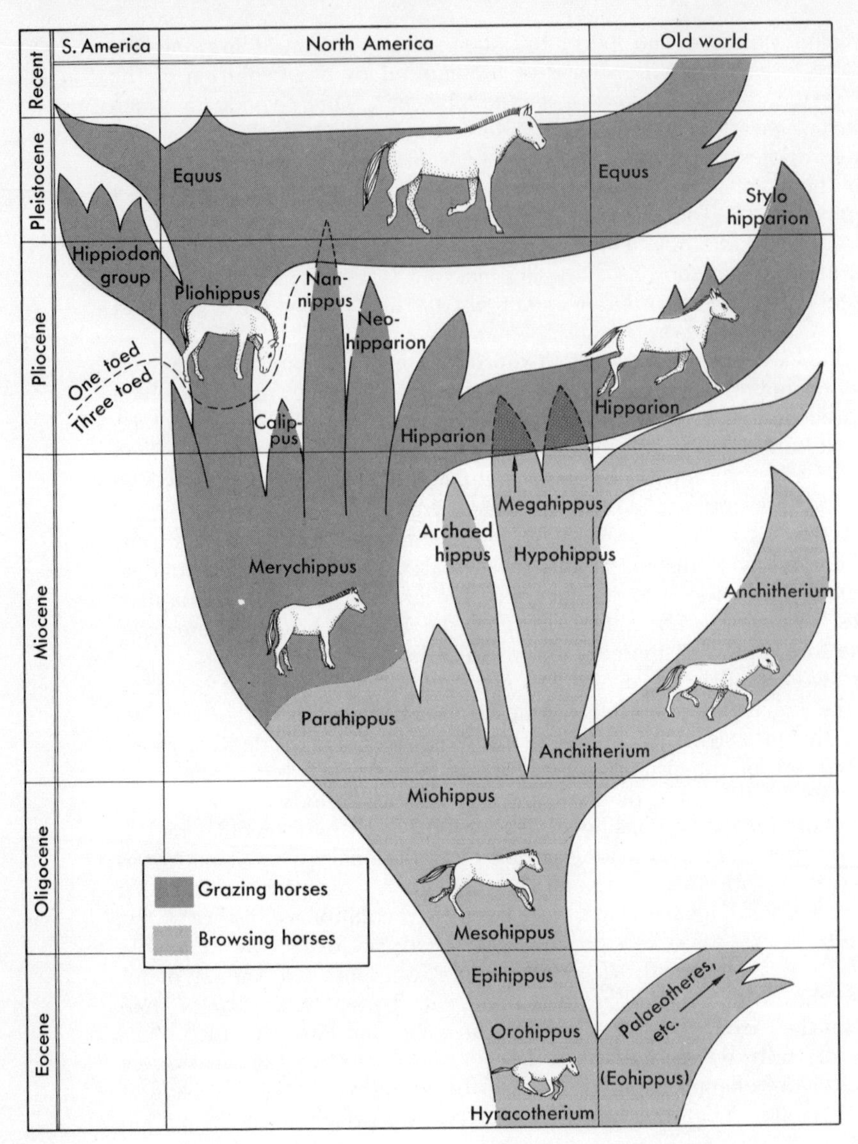

Fig. 7-7. The family tree of the horses. From Simpson, *Horses* (Oxford University Press).

hard, high crowned teeth equipped the later horses for grazing on tough, silica-containing leaves of grasses. To accommodate these teeth, the front part of the skull became deeper, and the face in front of the eye became longer.

4. Increase in size and complexity of the brain.

Contrary to the belief which was held by many paleontologists when relatively few fossil horses were known, these different trends of adaptive evolution did not progress at even and similar rates in all of the various branches of the horse line. As Figure 7-7 shows, horses underwent three major periods of adaptive radiation, one in the latter part of the browsing stage (early Miocene), another soon after the appearance of the first grazer (*Merychippus*), and a final one at about the time when the modern horse (*Equus*) appeared. At least the middle one of these radiations included a reversal of the usual trend in size, since a smaller animal (*Nannippus*) evolved from a larger one. The impression of "straight line evolution" for the group as a whole arises from the fact that these deviating genera of horses eventually became extinct. Each new successful radiation included both deviant side lines and one or more lines which progressed farther toward the modern horse.

Similar progressions can be followed in the evolution of many other groups of mammals. In some instances they involve the increase in size of such elaborate features as horns, antlers, and tusks. Such structures as the giant antlers of the extinct Irish deer and the tusks of some modern species of pigs (Figure 7-8) appear superficially to have become exaggerated to such an extent that they are no longer adaptive and might even be a detriment to the species. We must remember, however, that the species bearing these structures survived on the earth for thousands of years, and during this time probably produced millions of successful individuals. Because of the keen competition which exists in all habitats occupied by large mammals, these species could not have survived and been common for long periods of time if such structures were in any way detrimental. We cannot say what use they may have been to the animal. They may have developed because of the fact that males with the largest antlers, tusks, or horns had the greatest success in fighting off their rivals and mating with females.

The existence of such progressive trends has caused many evolutionists to believe that they are caused not by natural selection in a progressively changing environment, but by some internal guiding force. Such theories cannot be reconciled with the fact that these trends involve simultaneous changes in several different characteristics. Each individual characteristic, moreover, can be altered only by simultaneous changes in many different enzyme systems which are controlled by different, independently segregating

Fig. 7-8. Diagrams to show how lungs originated and evolved in fishes and primitive land vertebrates. From Romer, *The Vertebrate Story* (University of Chicago Press).

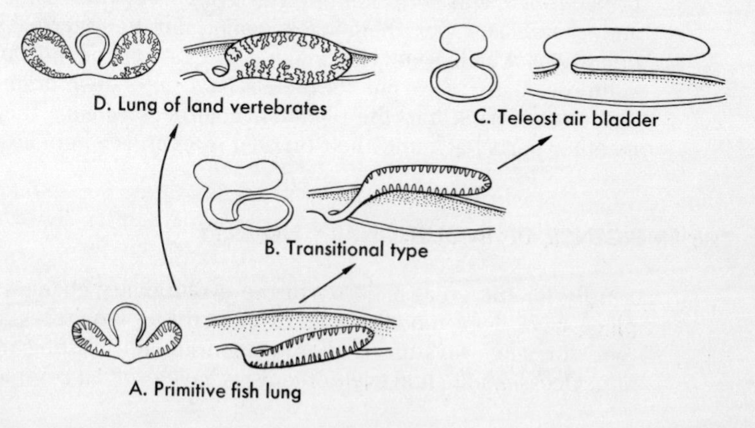

gene systems. Furthermore, as fossil evidence has accumulated, additional branch lines which are known to have deviated from the general trend, have always been discovered. The belief that progressive evolution is brought about by directed mutation or some unknown internal directive principle is rapidly passing into the limbo of discarded evolutionary theories.

Another evolutionary theory which has had to be modified as a result of more recent information is that of irreversibility. It is true that complex structures such as the gills of fishes, once they have been lost, cannot be regained in their original form. This is because the original evolution of these structures involved a complex sequence of precisely integrated genetic changes, guided by natural selection along a devious pathway which can never be retraced. Consequently, when terrestrial reptiles and mammals, which had lost all traces of gills, gave rise to secondarily aquatic forms such as the Jurassic ichthyosaurs and the modern whales, these animals developed completely different adaptations to life in the water.

Among plants, we can trace the evolution of petalless, wind pollinated flowers from ancestral types which had colored petals and were pollinated by insects. When later the wind-pollinated derivatives gave rise to insect or bird pollinated descendants, in some instances leaves became modified to serve the attractive function formerly assumed by petals. An example is the poinsettia, which we use as a Christmas decoration. The petal of a fully developed flower is a highly complex structure, with its own particular system of specialized epidermal cells, plastids, and veins, all of which are probably determined by different gene systems. Once it has become reduced and lost through a long series of evolutionary steps, its reconstitution by a similar series of steps requires a complex reorganization of the adaptive system. Modifying the developmental system of a leaf so that it can take over the attractive function of a petal requires far fewer genetic changes.

On the other hand, such relatively simple trends in adaptive characteristics as the increase in overall size of the organism or changes in the number of its similar parts, such as the number of vertebrae in the backbone of a reptile or of leaflets in the compound leaf of a clover or lupine, can be and often are reversed. Irreversibility is, therefore, not an intrinsic characteristic of evolution. Where it exists, it is a result of the complexity of the integrated sequence of processes which is responsible for any major evolutionary change.

By analogy, it is always possible to drive an automobile either forward or backward, and with skill one can even dismember some parts of its motor and chassis and put them back again. But to reverse all of the complex operations which went into making the car out of the raw materials which compose it, so as to put its metals back into their original ore, its rubber and plastics back into the plants or coal beds whence they came, and all of its other parts back into their original sources, is a hopelessly impossible task.

THE EMERGENCE OF EVOLUTIONARY NOVELTY

By far the great majority of the evolutionary changes which have taken place have been modifications of pre-existing structures and functional systems for new ways of life, based upon already existing methods of adaptation. Occasionally, however, organisms have evolved completely new adaptive systems, which have made possible new ways of life. Among vertebrate animals, the best known of these are the evolution of jaws and teeth in the earliest fishes; of lungs and walking limbs in amphibia; of large, hard-shelled eggs containing amniotic fluid in reptiles; of warm blood and the four chambered heart in mammals and birds; of hair, mammary glands, and the placenta in mammals; and finally of a thinking brain in man. Based upon the facts and theories which have already been presented in this book, we would not expect to find that the processes responsible for these major adaptive shifts are any different from those which operate to produce the more usual and less spectacular kinds of evolutionary changes. Nevertheless, they are of such importance to the course of evolution as a whole that we should pay particular attention to the way in which they have come about.

Unfortunately, most of the structures mentioned above are soft parts of the body, which are not preserved as fossils. We do have, nevertheless, good evidence to tell us how two major adaptive shifts took place in vertebrates. These are the evolution of lungs, and of the small bones, known as ossicles, which are found in the ears of mammals.

Lungs were not first evolved for life on land, but as a means of adaptive radiation by fishes into a particular kind of water. In stagnant, fresh water pools, the water can lose so much of its oxygen that fishes cannot use it as a source of oxygen for their gills. In such situations, fishes which can breathe only in this manner will die. There is, however, a group of modern fishes, known as lung fishes, which have two kinds of breathing apparatus. When they are swimming in normal water, well supplied with oxygen, they breathe through gills like those of other fishes. They are, however, adapted to living on the shallow margins of tropical lakes and streams, where in periods of drought the water becomes so foul and stagnant that other fishes die by the thousands. Under such conditions, the lungfishes rise to the surface and gulp in air. This enters a primitive lung, which opens into the digestive tract just below the throat (Figure 7-8). In this way, they can survive until the next wet season, when their water again becomes pure and well oxygenated. Consequently, lungs first evolved as a device by which a group of fishes was able to colonize an aquatic habitat into which their enemies could not penetrate, and where they had an abundant supply of food all to themselves.

Lung bearing fishes are first known from the beginning of the Devonian Period, 400 million years ago. They belonged to two closely related groups, one of which contained the direct ancestors of modern lungfishes, and the other the ancestors of amphibians. Still others were the ancestors of modern bony fishes. From the primitive lung bearing fishes there evolved one line which returned to deeper water which was always well aerated. In these fishes, the lung became converted into a dorsal swim bladder. This organ is present in all modern bony fishes, and greatly helps them in swimming, by adjusting their hydrostatic pressure to different depths of the water. In some primitive modern fishes, such as the gar pike (*Lepidosteus*) and bowfin (*Amia*), found in lakes and streams of the central United States, the swim bladder can still function as a lung when the water becomes polluted.

Another group of lung bearing fishes, the lobe fins, had strong bony fins which evolved into primitive legs, as already mentioned. During all of the period of evolution from the lobe fin stage to that of primitive amphibians, which extended over about 50 million years, the wholly or chiefly

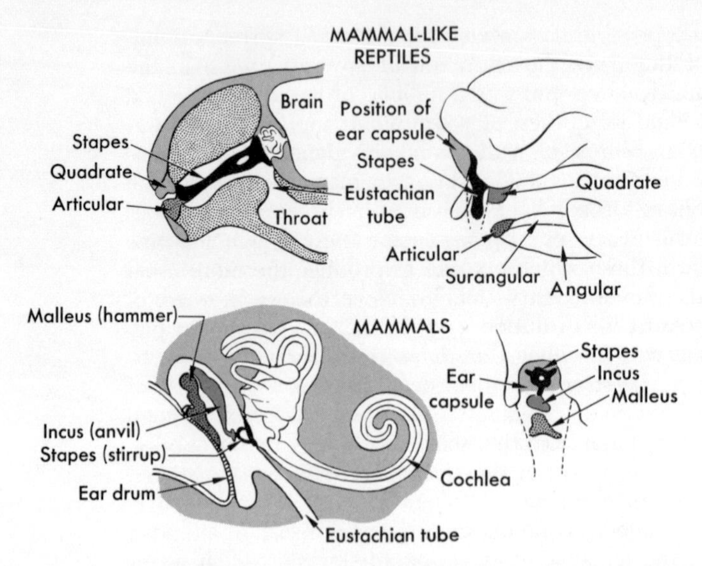

Fig. 7-9. Diagrams to show the origin of the bones in the ear of mammals, from certain bones of the reptilian jaw. From Romer, *Vertebrate Paleontology* (University of Chicago Press).

aquatic animals constituting the line which led to land vertebrates were equipped with two kinds of breathing apparatus, gills for well aerated water, and lungs for air breathing when the water was polluted. Most lines of amphibians then took to living the greater part of their adult life on land, and consequently lost their gills.

This course of evolution illustrates the two principal ways by means of which major adaptive shifts can be brought about. One is the existence during the transitional period of TWO KINDS OF ADAPTATIONS FOR THE SAME GENERAL FUNCTION, such as gills and lungs in lungfishes and primitive amphibians. The other is the presence of an organ which AT THE SAME TIME HAS TWO FUNCTIONS, like the primitive lung and swim bladder of many early bony fishes.

Another example of a series of structures which had double functions during a period of transition is that of the bones in the mammalian ear (Figure 7-9). In all mammals, including ourselves, the ear apparatus contains four specialized bones. One of these surrounds and reinforces the ear drum, while the other three amplify the sound waves and transmit them from the ear drum to the auditory nerve in the inner ear. In reptiles, the ear contains only one of these bones, which corresponds to the innermost, or stirrup, in the mammalian ear. The other three mammalian ear bones have direct counterparts in bones of the reptilian jaws, one belonging to the upper and two to the lower jaw. In the transitional mammal-like reptiles, described earlier in this chapter, these bones had become very small, and in some instances were loosely attached to the jaw. Moreover, some of these animals had moderately well-developed extra ear drums opposite these jaw bones. In all probability, therefore, these jaw bones aided the hearing function by their vibration even before they became separated from the rest of the jaw. Therapsid reptiles having this condition represented the line leading to mammals for most of the Triassic Period, or about 40 million years.

EVOLUTIONARY CONSTANCY

When an evolutionary line has gone as far as it can along a particular line of adaptive radiation, it can do one of three things. If the environment opens up new evolutionary opportunities, it can start on a new course of diversification and adaptive radiation. If the conditions to which it is adapted become changed or greatly reduced in extent, and if other organisms have evolved which can take better advantage of these restricted conditions, it will become extinct. If, on the other hand, the environment to which it is adapted remains constant for long periods of time, and does not change significantly with respect to the needs of the particular organisms composing it, a line of organisms can remain in a stable equilibrium with their environment for indefinite periods of time, without showing any signs of evolutionary change.

There are many examples of such evolutionary stability, which have lasted in some instances for hundreds of millions of years. Some of them are illustrated in Figure 7-10. There is no reason for believing that this stability is brought about either by reduction of the size of the gene pool to such an extent that genetic variation is impossible, or by a drastic curtailing of the

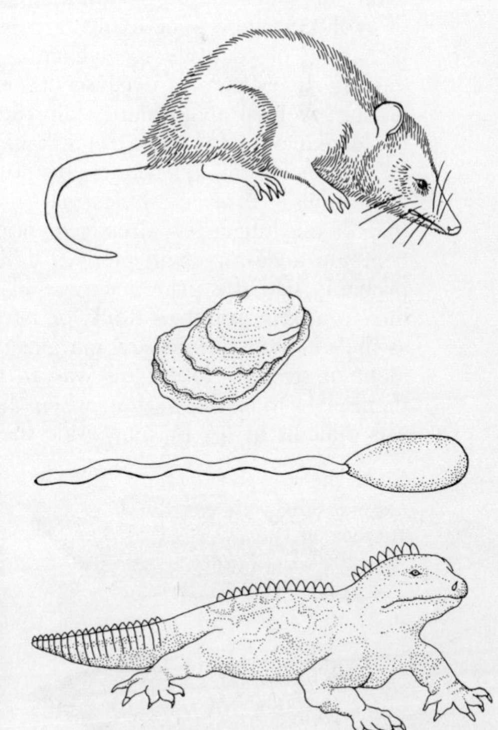

Fig. 7-10. Some modern animals which are the products of varying degrees of stabilization in evolution. *Top,* the opossum, which, compared to other modern mammals, evolved relatively little from marsupial ancestors which lived 80 million years ago. *Middle,* the oyster (*Ostrea*) and the lamp shell (*Lingula*), which are hardly different from their ancestors which lived, respectively, 350 and 500 million years ago. *Bottom,* the tuatara (*Sphenodon*), a lizard-like reptile, the only survivor of an order which flourished during the age of dinosaurs, 150 million years ago. From Simpson, *The Meaning of Evolution* (Yale University Press).

mutation rate. One of the best known mammals which, compared to other modern mammals, has changed relatively little during the past 60 million years is the opossum (*Didelphis virginiana*). The phenotypic variation between individuals in modern populations of opossums has been measured and is greater than that of wood mice and other rodents living in the same region. These rodents belong to rapidly evolving lines of evolution. Moreover, opossums have successfully invaded California in company with the white man. There they have now become quite common, and have amply demonstrated their ability to cope with the new climate and man-made environment which exists in the more highly populated parts of this state. An example of evolutionary stability in plants is the giant Sequoia (*Sequoiadendron giganteum*), which has remained stable for tens of millions of years, and has greatly reduced its area of distribution. Giant sequoias, introduced during the past hundred years into Great Britain, where the climate is very different from that in their Californian home, have proved to be highly successful as cultivated trees. Moreover, nurserymen have been able to select a great variety of unusual genetic types from the seed stocks at their disposal. Since giant Sequoias produce their first seeds at so great an age that no cultivated trees are more than one generation removed from their wild ancestors, the great majority of the genes responsible for the selected variants of this species must have been present in its native gene pool.

In some instances, lines of evolution which have been stabilized for long ages possess ancient fossil lineages which reveal the earlier stages of their evolution. In this way, paleontologists have learned that these constant lines of evolution were once evolving rapidly, and had, for their time, reached a high level of adaptive specialization. They appear primitive and archaic in modern times merely because other lines of evolution have reached even higher levels of adaptation while they have remained static. What, then, is the explanation for these remarkable examples of evolutionary constancy?

Although this question is by no means answered to the satisfaction of evolutionists, three examples may provide a clue to an answer. The first is that of the lungfishes, already mentioned several times in this chapter. The common ancestor of all modern fishes except for sharks and their relatives probably had the primitive type of lung already described. One radiating line from this primitive stock, or perhaps two closely related lines, adopted as their habitat the shallow margins of lakes and streams. Here the development of strongly lobed fins was an adaptive advantage, either for pushing themselves over the bottom when the water was so shallow that swimming was difficult or for pushing aside the heavy vegetation which choked these

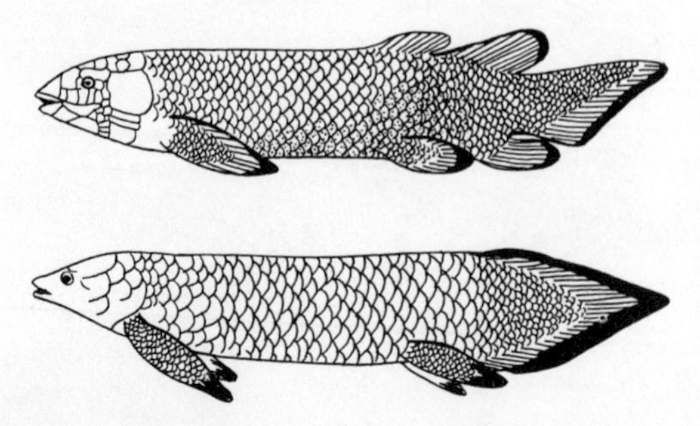

Fig. 7-11. A primitive lungfish (*Dipterus*), which flourished during the Devonian Period, and its modern descendant (*Epiceratodus*), found in South Australia. During the Triassic Period there existed lung fishes (*Ceratodus*) which had tooth plates and head bones very much like those of the modern Australian species. Complete skeletons of these Triassic lungfishes have not yet been found. From Colbert, *Evolution of the Vertebrates* (John Wiley & Sons, Inc.).

shallow waters. These two lines then diverged in their diet. One of them, as shown by their teeth (Figure 7-12C,D), retained their ancestral habit of actively pursuing quickly moving prey. These became the lobe-fins, which gave rise to amphibians and then became extinct, apparently because of unfavorable competition with their more efficient descendants. In the other line, the teeth became greatly modified for grinding both vegetation and shellfish (Figure 7-12E,F). As these teeth evolved, the scales of their bodies became thinner, the fins became reduced and weaker, and the fish acquired all of the bodily features associated with sluggishness. This evolution occurred comparatively rapidly, so that in about 25 million years fishes which were not unlike those in Figure 7-2 had given rise to the type shown in Figure 7-11A. From then on evolution was slower, and 140 million years later a fish had

Fig. 7-12. Lower jaws and teeth of lobe fins (A, B, C, D) and lungfishes (E, F), to show how divergence in teeth and diet led, on the one hand, to evolution in the direction of land life, and, on the other hand, to evolutionary stabilization. Further explanation in the text. From Romer, *Vertebrate Paleontology* (University of Chicago Press).

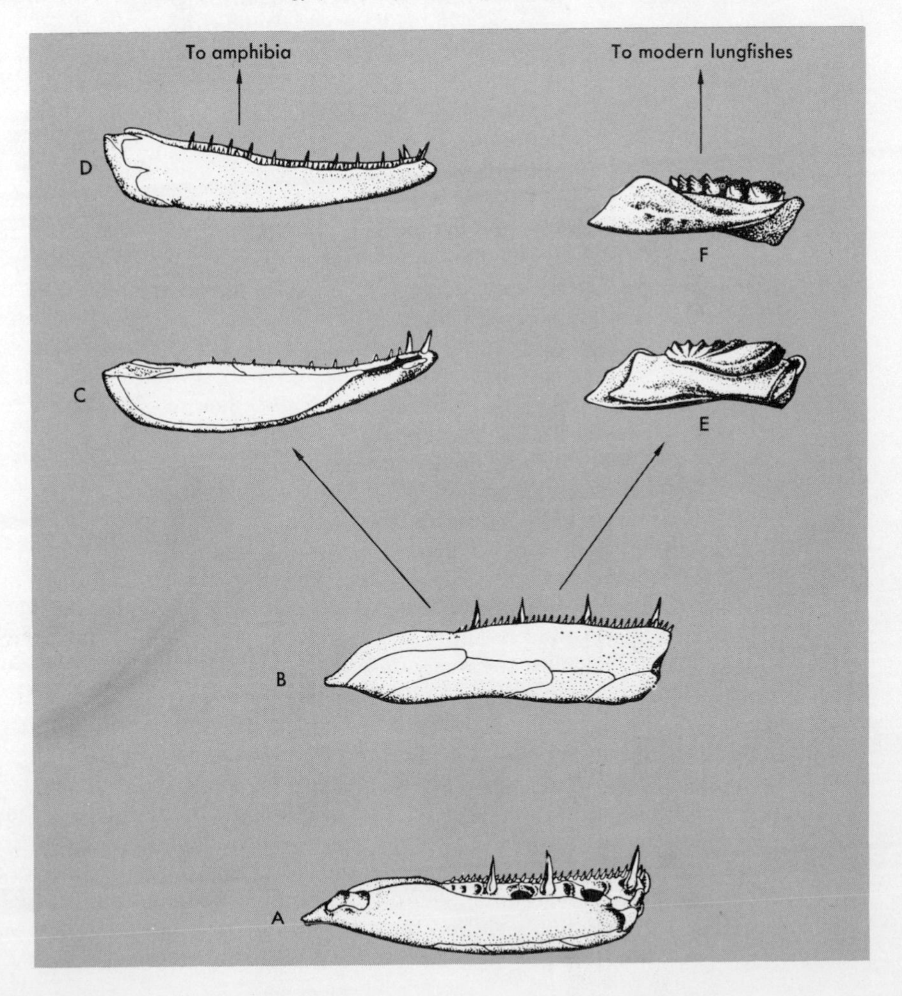

evolved which was very much like the modern Australian lungfish (Figure 7-11B). Since this Triassic lungfish lived about 200 million years ago, the line has been constant for a very long period of time. Two other species of modern lungfishes, one in central Africa and one in South America, have evolved even farther toward degeneration of fins and sluggishness, but these also have very similar ancestors which lived about 150 million years ago.

In this case, constancy appears to be the result of life in a specialized habitat, which predators have difficulty entering, and of perfect adaptation to feeding on a generalized type of food which has always been easily available.

The second example is the Virginia opossum. This mammal inhabits deciduous forests, a habitat which has been widespread and continuously available ever since the adaptive radiation of the early marsupials gave rise to its first opossum-like derivative. It is sluggish in habits, much less intelligent than the placental mammals which inhabit the same forests, and has only two methods of defense against more powerful carnivorous predators. One is escape by "playing 'possum," which means feigning death to such a degree that many predators do not attack it. The other is a high rate of fertility. In addition, the diet of the opossum is very varied: small slow-moving insects, grubs, seeds, and anything else which can serve as nourishment. Its methods of escape and of food gathering are both so generalized that they have been effective in relation to all of the evolutionary changes in species of predators and of food animals which have taken place in the forest habitat over millions of years.

The third example is the tuatara (*Sphenodon*), a remarkable reptile which looks like a lizard, the only survivor of an order (Rhynchocephalia) which flourished at the time when the dinosaurs were beginning their evolution, and which has evolved little since the Jurassic Period, 150 million years ago (Figure 7-11B). This reptile feeds on a generalized diet of insects, and has escaped more efficient predators by its isolation on New Zealand. These islands have been cut off from the mainland since before either the dinosaurs or mammals had reached their peak, and were never inhabited by representatives of either of these groups until mammals were introduced by human colonizers. The tuatara survives in modern times only because it exists on a small island south of the principal South Island, and is carefully preserved by the New Zealand Government.

The common feature of these three constant lines is that they are all adapted to a very generalized diet, which has been continuously available. All have escaped from predators in ways which, though different, have been equally effective against the many different kinds of predatory animals which have evolved during tens or hundreds of millions of years. Each additional example of evolutionary stability which might be cited would differ from these three in various ways, but their common denominator can be expressed as follows. Populations of organisms rarely if ever have significant interactions with all of the factors of their environment. Furthermore, some environments have remained essentially constant over long periods of time in respect to many of the significant factors for maintaining certain kinds of organisms, even though these environments may have become much reduced in extent, or have slowly shifted their geographic positions from one part of the world to another. Some evolutionary lines have become well adapted to these unchanging factors of their environment. They have, therefore, maintained a stable equilibrium for millions of years. Mutations and new gene recombinations have been constantly occurring in them, but have with equal constancy been removed by stabilizing selection.

Many groups of animals have retained evolutionary stability through stabilizing selection connected with two passive methods of defense against predators, extreme fecundity and protective armor. A fine account of them is given by the Russian evolutionist, I. I. Schmalhausen, in his book, *Factors of Evolution*. They include groups of radiolarian Protozoa; sponges; some kinds of jellyfishes; certain annelid worms; simple small crustaceans related to shrimps; various groups of shellfish or molluscs; the king crab (*Limulus*), common along the Atlantic Coast of North America; and the cockroaches. Evolutionary constancy for tens or hundreds of millions of years is not an exceptional or unexpected condition. It is the normal result of certain kinds of interactions between populations and various relatively constant types of environments.

This consideration of evolutionary constancy emphasizes the concept of evolutionary opportunism, which has emerged from all of the facts and theories presented in the preceding chapters. In order to survive and evolve, populations of organisms need a pool of genetic variability which enables them to establish successful relationships with certain factors of their immediate environment. Depending both on the nature of the organism-environment relationship present at one time, and on the way in which the environment changes relative to the qualities and potentialities present in the available gene pool, an evolutionary line of populations may progress either rapidly or slowly, may become extinct, or may remain constant for long periods of time. There is no evidence in biological evolution of a grand overall design or of any kind of predestination.

Chapter Summary

The time scale of evolution as a whole is so much greater than that of evolutionary processes which can be observed by men under contemporary conditions that events which in the fossil record appear to have taken place almost instantaneously may have been in terms of population change per generation so slow as to be almost imperceptible. Fossils are now dated on the basis of decay of various radioactive materials, particularly carbon-14, uranium-238, and potassium-argon.

If genera, families, and other higher categories are relatively old, and if many intermediate forms have become extinct, they are well marked and can be delimited in a reasonably objective fashion. In actively evolving groups, in which little extinction has occurred, these categories are very hard to define, and their definition is more or less subjective. Individual characteristics which in some groups of organisms are diagnostic of families are in other groups useful only for separating genera or species. These facts suggest that the evolution of higher categories has been by means of the same processes which have brought about the evolution of races and species, except that extinction of intermediate forms has played an increasingly important role as the higher categories have become older.

The fossil record of vertebrates includes forms intermediate between fishes and amphibia, between amphibians and reptiles, between reptiles and mammals, and between reptiles and birds. It suggests that in certain past ages the characteristics which now distinguish these five classes were separating groups which from the point of view of genetic relationships were no farther apart than are present day genera or families.

Progressive evolution in the vertebrates was always accompanied by adaptive radiation and is best characterized as a succession of adaptive radiations. A classic example is the evolution of the horse. Modern interpretations of the fossil record do not support the idea that progressive evolution was caused by internally directed trends or "orthogenesis," which were independent of adaptation and selection, and which produced characteristics so much exaggerated that they were inadaptive. Irreversibility in evolution is true only for major trends of which the separate components are so numerous and carefully integrated that the complex sequence of steps which has occurred cannot be traced backward.

New characteristics appear in evolution chiefly in two ways: 1. During an intermediate, transitional stage a single organ has two functions, or 2. two organs serve the same function. Examples of 1. are the evolution of the swim bladder in the more advanced fishes and the small bones or ossicles in the mammalian ear. An example of 2. is the evolution of lung breathing from gill breathing in amphibians.

Many lines of evolution have remained essentially constant for millions of years. This constancy is associated with a long term equilibrium between the population and its environment, such that only stabilizing selection occurs, constantly rejecting new mutations and gene combinations as they take place. In animals this condition has been most often associated with very generalized methods of defense against predators, such as sluggishness, heavy armor, and unpalatability, and particularly high fecundity.

Questions for Thought and Discussion

1. Give as many reasons as you can why the fossil record of vertebrates is more complete and easier to interpret than is that of either higher plants or insects.

2. Fossils of molluscs (shellfish) are far more common than are those of reptiles and mammals; yet except for a few particular trends the evolutionary history of the molluscs is much harder to interpret than is that of reptiles and mammals. Explain why this is so.

3. Discuss possible reasons why species or families of animals become extinct. Why is extinction likely to make genera and families easier to define?

4. What are the connections between adaptive radiation and progressive evolution?

5. Based upon your knowledge of genes and their action, give your opinion of the chances that structures will be altered progressively by internally directed forces, operating independently of the environment and of natural selection.

Processes of evolution in man

CHAPTER 8

In the final chapter of this book, we shall see how the concepts developed in previous chapters apply to the evolution of our own species. According to them we should expect to find in man's evolutionary past a series of populations equipped with a pool of genetic variability evolving through time under the guidance of natural selection. They should have undergone adaptive radiation in response to significant changes in the environment, both at the level of differentiating races within species, and of the origin of new and different species. To a certain extent, this is exactly what many paleontologists and anthropologists now realize that they have found.

MAN'S POSITION IN THE ANIMAL KINGDOM At the outset, two points must be emphasized. First, man is unquestionably an animal in his origin and biological characteristics. Anatomically, he is much like the large apes, particularly the chimpanzee and gorilla. Moreover, recent comparisons have been made between the chemical structure of human proteins, particularly the hemoglobins of our red blood cells and the proteins of our blood serum and the corresponding proteins of apes. These comparisons have shown that men, chimpanzees, and gorillas bear a closer resemblance to each other in respect to these basic chemical substances of the body than any of them does to other apes, such as the orangutan and gibbon. This evidence strongly favors the hypothesis that at one stage of their evolution the ancestors of man were apes, which had already been differentiated along a line of adaptive radiation different from that which led to the orang and gibbon, and which radiated again to give rise to the gorilla, chimpanzee, and man. As will be described below, this hypothesis is entirely in accord with the most recent fossil evidence.

Nevertheless, man is more than a biological animal. Looking at our own species from the most objective viewpoint possible, we can recognize our unique character. No other species can control its destiny, to the extent that we do, by our ability to remember and profit from the past, to look ahead and imagine the future, and to talk with each other and work together toward achieving a better way of life. Furthermore, these three abilities, once acquired, have enabled man to change his way of life at a pace many times faster than that which any other species has achieved and in a qualitatively different way. He has done this during the past thirty-five thousand years without undergoing any recognizable changes in his anatomical features or brain power. Consequently, we cannot fully understand human evolution unless we consider both its purely biological as well as its cultural aspects and the interrelation between them.

THE FOSSIL RECORD OF MAN

More than a half century ago, in 1894, a young Dutch anatomist, E. Dubois, electrified the scientific world by reporting that he had discovered the "missing link." The German anatomist Haeckel, a follower of Darwin, had predicted that, on the basis of evolutionary theory, fossils would be found of creatures intermediate between apes and men. Dr. Dubois asked to be appointed as army doctor in the East Indies so that he could search for such fossils. In river deposits at Trinil, in eastern Java, he found first a skull cap, then a femur, and other remains.

The experts of Europe and America agreed on the importance of this discovery, but disagreed on almost everything else. Some said that it was just an ancient man with an abnormally low forehead. Others proclaimed it to be an ape related to the orangutan or perhaps the gibbon or chimpanzee. Since Java man started the search for fossils which would reveal man's ancestry, the scientific world has been a succession of discoveries of fragmentary bones belonging to man's evolutionary line, accompanied by a corresponding succession of arguments about their nature and relationships.

During the past twenty years, the number of known fossils of man and his ancestors has greatly increased. Furthermore, both these newer finds and the older ones are being interpreted in a different way. In the past, the scientists who described these fossils were thinking in terms of individual types rather than populations. They asked themselves: Is this newly found fossil different in any way from other known individual fossils? If the new find proved to be different, and it nearly always did, they gave it a scientific name, placing it in a different species and often a different genus from all other human or man-like fossils. We now realize that the differences between many fossils found in strata of the same or similar age are no greater than those between, for example, an Australian bushman and a tall tribesman from East Africa, a slender Amazonian Indian and a stocky Eskimo, or a narrow faced man from the Middle East and a broad, round headed central European. Consequently modern students of these fossils, both anthropologists and zoologists, are tending to discard nearly all of the names of "genera" which have been erected in the past. They recognize that since their divergence from the apes, the ancestors of man have progressed chiefly along a single line of evolution. At times this line has branched to produce two or three related and sympatric species, but during at least the past 600,000 years it has probably consisted of a single species, possessing a common gene pool, and subdivided into a number of different races. For these reasons, every account of human evolution written before 1950 is already or will soon be obsolete.

Unfortunately, the fossil record of man's ancestry, though much better than it was even a few years ago, is still very incomplete. We can begin with a group of apes which were common in Africa and Asia during the Miocene Epoch, from about 25,000,000 to about 13,000,000 years ago, and are collectively designated dryopithecines. They undoubtedly consisted of several species, and are usually placed in a number of different genera. Some of them inhabited forests and were tree climbers like the modern apes, while others apparently lived in open savannas and walked about on all fours. Few if any of them had developed fully the specialization of modern apes, such as their very long arms by which they move hand over hand from one tree branch to another; the "simian shelf" of their jaws, which supports strong muscles used for tearing and chewing the bark of trees; and their large canine teeth with which they crush hard nuts.

Since only a single skull belonging to one species of these apes is preserved in its entirety, good estimates of the size of their brains are not available. Probably, they were somewhat smaller than those of modern apes. These apes were sufficiently generalized so that they could have formed the ancestral stock from which the line leading to man arose.

From the end of the Miocene to the end of the Pliocene Epoch stretches a gap of more than ten million years, from which very few fossils belonging to the line leading to man are known. At the end of this gap we find two separate kinds of creatures which may have been forerunners of man. One of these was that of the australopithecines or "southern apes," so named because their first remains were found in South Africa. More recently, abundant remains of australopithecines have been found in East Africa, and a few fragments of teeth and jaws which may belong to members of this group are known from Java.

In respect to their intelligence and their way of life, these animals were still more like apes than men. Their brains were only slightly larger than that of the chimpanzee, and hardly more than one-third the size of the brain of modern man. They did not use fire, probably did not build shelters, and in all probability could communicate with each other only by means of crude cries or grunts. On the other hand, they did walk erect, and they probably used crude tools. Their limb and body skeletons are much like those of modern man. Their skulls show a mixture of ape-like and man-like characteristics. Their large jaws and small brains ally them to the apes, while in their relatively small, regular front teeth they are much like men.

There are still many uncertainties about the exact age of many of the fossil australopithecines. Present finds and their dating suggest that this group of apes existed from at least the later part of the Pliocene epoch (two million years ago) until well into the Pleistocene or ice age, perhaps 700,000 years ago. There were at least two species, which in some places were sympatric and did not intergrade. One was a larger animal, which had relatively large

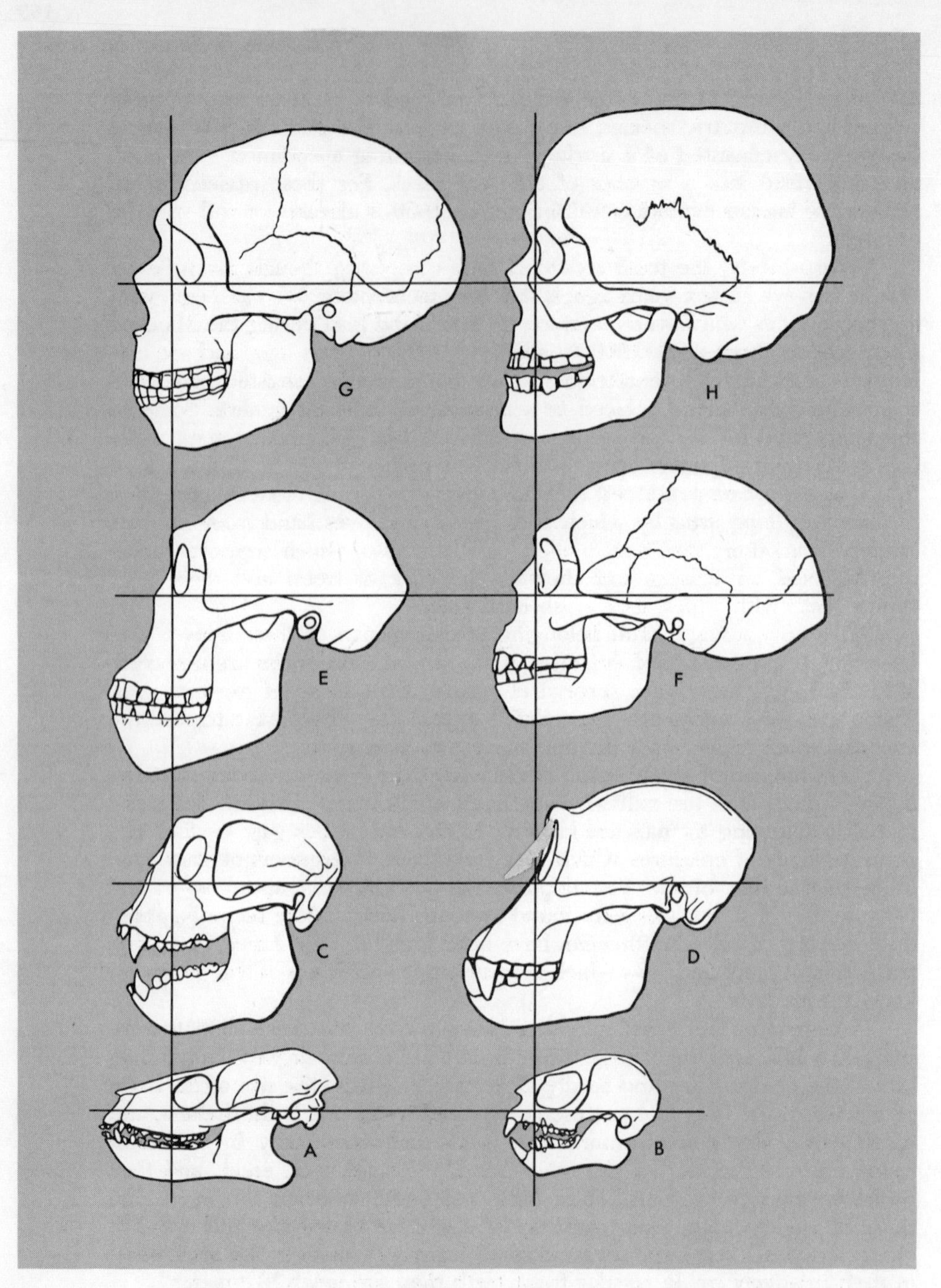

Fig. 8-1. Skulls of various primates showing changes in size and proportion during their evolution toward modern man. *Bottom row,* (A) a lemur-like and (B) a tarsier-like animal, both of which lived during the Eocene Epoch. *Second row from bottom,* an extinct Old World monkey (C) and a modern chimpanzee (D). *Upper middle row,* An Australopithecine (E) and "*Pithecanthropus*" or *Homo erectus* (F). *Top row,* Neanderthal man (G) and Crô-Magnon man (H), both of which are now placed by many biologists and anthropologists in our own species (*Homo sapiens*). From Colbert, *Evolution of the Vertebrates* (John Wiley & Sons, Inc.).

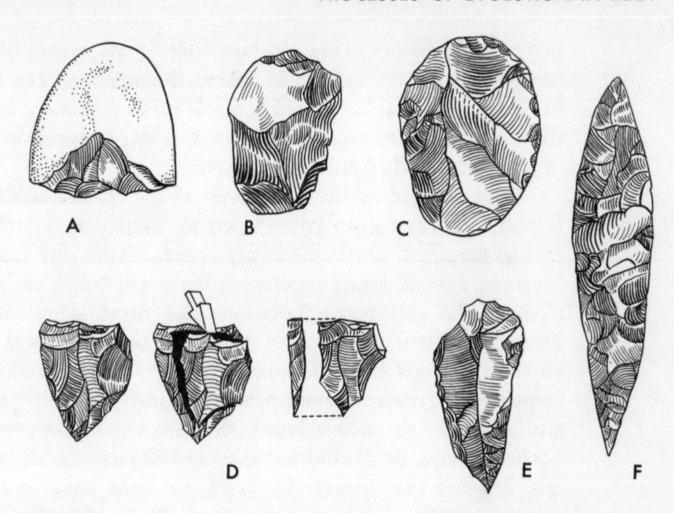

Fig. 8-2. Representative tools of prehistoric men and their ancestors. (A) A pebble tool from the earliest Pleistocene, similar to those which have been found in association with Australopithecines. (B) A crude but more effective chopping tool found in the cave from which the remains of Peking man (*Homo erectus*) were obtained. (C) A much better fashioned hand-axe from middle Pleistocene deposits of Europe (Acheulian), perhaps fashioned by more advanced men of the species *Homo erectus*. (D) A later tool made by knocking a flake from a previously fashioned stone, from the period (Levalloisian) of the upper Pleistocene which has also yielded fragmentary bones approaching the form of modern man. (E) A pointed flake tool of a later date (Mousterian) and finer workmanship. This type is associated with the remains of Neanderthal man. (F) An exquisitely fashioned blade associated with a culture which flourished briefly in Europe about 20,000 years ago (Solutrean), and was definitely associated with modern man. From Howells, *Back of History* (Doubleday & Company, Inc.).

teeth and may have been strictly vegetarian in diet. The other was smaller but had a brain just as big or slightly bigger than that of its larger relative. Judging from the appearance of its teeth, it probably ate some meat and perhaps killed small animals.

Very recently, in the same beds as those which contain the East African australopithecines, a small number of fossils has been found of creatures which were exactly intermediate between australopithecines and the most primitive species of man (*Homo erectus*), to be described below. This intermediate form, which has been given the name *Homo habilis*, was probably the maker of the crudely chipped stone tools which are found in association with both its fossils and those of some australopithecines (Figure 8-2A). The discovery of *Homo habilis*, which was described as recently as 1964, tells us two things. In the first place, the known fossils of australopithecines do not represent the direct ancestors of man, since they lived contemporaneously with the more advanced *H. habilis*. Nevertheless, the evolutionary line which led to man most probably passed through a stage similar to the known australopithecines. There is every reason to believe when fossils are discovered belonging to representatives of man's evolutionary line which lived during the earlier and now unrepresented Pliocene Epoch, they will be like the known Pleistocene australopithecines. Such discoveries would tell us that australopithecines were, in fact, direct ancestors of man. They nevertheless persisted side by side with their derivatives, the earliest men, for several hundred thousands of years. Such contemporary existence of related primitive and more advanced species is a very common situation in other groups of animals, both fossil and modern.

The second message conveyed by *Homo habilis* is that, as Darwin believed, the transition from apes to man was a truly gradual one. The australopithecines were certainly apes; they did not possess any distinctive features absent from modern apes except the ability to walk erect. Nevertheless, the difference between the most advanced australopithecines and *Homo habilis* is no greater than that between two closely related species of animals, except that the stone tools probably made by *H. habilis* must have required a greater development of the learning process than that found in any ape. On the other hand, the differences between *Homo habilis* and the earliest forms of *H. erectus* (see below) are in all respects only quantitative, and are not very great. In short, we now have available a series of fossils—dryopithecines, australopithecines, *Homo habilis, H. erectus, H. sapiens*—which forms a complete and gradual transition from apes to modern man.

In the early middle Pleistocene directly following the time when the last *H. habilis* existed, remains of creatures are found which are now generally regarded as belonging to the genus *Homo.* They include the original fossils discovered by Dr. Dubois plus several more recently discovered remains of the same age and kind from Java. They also include Peking man, discovered in caves near Peking, China, as well as isolated fragments from other parts of the Old World, such as Africa, and perhaps the jaw discovered long ago near Heidelberg, Germany. These fossils are all similar enough to each other so that they could have belonged to the different races of a single species, but they are so different from modern man that they deserve rank as a separate species, *Homo erectus* (Figure 8-1F).

The body skeleton of *Homo erectus* was essentially the same as that of modern man. He differed mainly in his massive skull, larger teeth, and smaller brain. The average brain size of this species of man was about 75 per cent as large as that of modern man and twice as large as the brain of the australopithecines.

The fossils of these earliest men are often accompanied by a large quantity of stone tools, including hand axes, which they must have made and

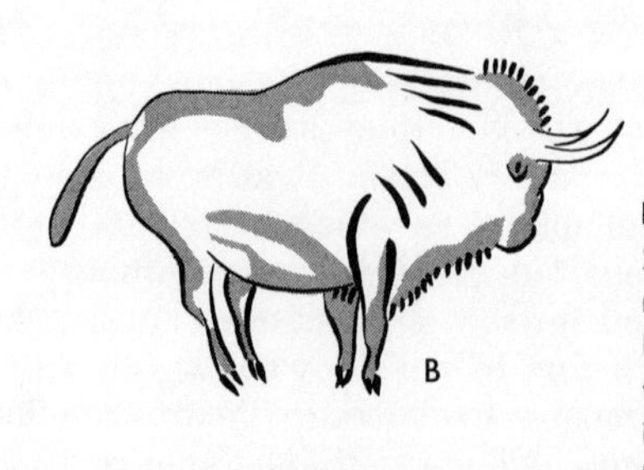

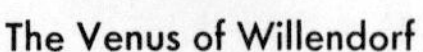
The Venus of Willendorf

Fig. 8-3. Artistry of the men who lived 20,000 to 25,000 years ago. (A) A renowned statuette, the "Venus of Willendorf". (B) A bison from the ceiling of the Altamira cave in Spain. The original is painted in red, yellow, and black. From Howells, *Back of History* (Doubleday & Co., Inc.).

used. They differ from the crude, chipped stones used by *Homo habilis* in having a series of definite designs, which were repeated in various localities throughout Eurasia and Africa, and showed a steady progression of improvement through time (Figure 8-2B,C). It is hard to see how men could have made such well-fashioned tools, apparently according to established traditions, unless they were able to teach each other how to make them. We can suppose, therefore, that *Homo erectus* possessed at least a primitive form of speech. In all of his fossil sites, moreover, are found bones of large animals such as giant pigs, sheep, oxen, baboons, horses, hippos, and elephants, which these hunters apparently killed. To do so they must have hunted in well organized bands, which suggests that they already possessed a well developed tribal structure. Some of the sites contain charred pieces of wood, indicating that their occupants used fire.

Recent interpretations of the fossil record by some paleontologists, as well as by zoologists acquainted with modern species of animals, have led to the belief that from the time of *Homo erectus* (600,000 years ago) to the first appearance of modern man, the entire habitable land masses of Eurasia and Africa were apparently occupied by a single species of man. Through the ages this species evolved gradually and irregularly in brain size, certain anatomical features, and culture. At all times it was subdivided into many races, which arose in various places, sometimes spread to other parts of the earth, and from time to time either eliminated other races by conquering and killing them or combined with them through interracial mating. The first men which were anatomically indistinguishable from ourselves appeared during the last advance of the glaciers, from thirty-five thousand to forty thousand years ago. Their appearance was accompanied and followed by a rapid expansion, diversification, and improvement of culture. People of this period made elaborate, beautifully fashioned stone axes and spears which modern man could not imitate without years of practice (Figure 8-2D,E). They carved figures of animals and people, and made lifelike paintings on the walls of their caves (Figure 8-3). Since they buried their dead together with implements carefully laid around the body of the deceased, we can imagine that they had some kind of religion and believed in an after life. The age of modern man, *Homo sapiens,* had begun.

POPULATION STRUCTURE AND SELECTION IN HUMAN EVOLUTION

We have no way of knowing anything about the variability within populations of the ancestral apes and australopithecines, and the fossil record is still too incomplete for us to say to what degree they were differentiated into distinct species. As already suggested, however, the present evidence suggests that since *Homo erectus* evolved about six hundred million years ago, mankind has possessed a common gene pool. If this is true, then man's ability to wander over long distances would enable the bearers of any improved genetic characteristics to transmit them throughout most of the range of this species in a relatively short time, at most a few hundred years. On this basis, it is idle to speculate on just where or when a particular trait or biological character complex first appeared. The evolution of man from *Homo erectus* to our modern species should not be visualized in terms of the origin of particular types at certain places and their subsequent spread as distinct entities. We should, rather, imagine that various new and valuable traits appeared in different parts of man's vastly extended range of distribution. People who became successful because they had newly acquired a particular trait subsequently migrated to different parts of the world, and by conquest, intermating, or both, transmitted this trait to other races of people. Another different trait could have originated in an entirely different part of the world and could have combined with the first one in a race inhabiting still a third region. The pieces of man's mosaic evolution may have been fashioned separately in regions far apart from each other. They may have been assembled and reassembled in different combinations at different places and times, until finally a particularly successful combination lifted man's evolutionary line up to a new level of adaptation.

The evidence now available, fragmentary though it is, can best be explained by the hypothesis that mankind has always been subdivided into races. Their evolution has included both divergence in isolated regions and fusion of pre-existing races through intermating. Races have likewise become extinct both through being conquered by other, more efficient races, and by genetic mixing. Consequently, there is no need for speculating on what may have been the relationships between the races of modern man and those which existed 25,000 years ago or earlier. For instance, during historic and late prehistoric times western Europe was repeatedly invaded by people coming from the east, who eliminated most of the older inhabitants, such as the Cro-Magnon men who made the famous cave paintings of France and Spain (Figure 8-3). Nevertheless, anthropologists have good reasons for believing that the Basque people of northwestern Spain and the Berbers of the Atlas Mountains in North Africa may be the relatively unaltered descendants of the Cro-Magnon stock. If so, the fact that these people have frequently intermarried with descendants of the invaders indicates that all modern Europeans and their American descendants contain many genes derived from the Cro-Magnon people. At an earlier time, the heavy-bodied, brutish-looking, but relatively highly cultured people of the ice age known as Neanderthals, were apparently extinguished by conquest in western Europe. On the other hand, in caves near Carmel, in Israel, there is evidence that Neanderthal-like people intermingled and exchanged genes with people belonging to a race similar to the Cro-Magnons. Such mixing, however, was probably not confined to the caves where the intermediate individuals have been found. Knowing the habits of conquering men, one finds it difficult to imagine that the Neanderthals were killed off by men of other races before their females had made a considerable contribution of genes, both voluntary and involuntary, to the conquering races.

In Africa, where the ice age did not decimate the population, but on the other hand rendered habitable large parts of the Sahara desert because of the increased rainfall, the divergence and mixing of races was, if anything, greater than in Eurasia. Some African races, like the South African bushmen, are apparently very ancient. Others, like the Negroes, are very recent and probably of mixed origin. The recent races of contemporary man did not, however, evolve from primitive, extinct species of man but from pre-existing races of *Homo sapiens*. Many of the biological features of Negroes, for instance, are more advanced than are those of any other races, in that they are more divergent from the features of primitive man and his ancestors. These advanced traits include thick lips, curly hair on the head, and the lack or scarcity of body hair.

These facts are mentioned in order to point out that from the point of view of the evolutionist, the term "purity of the race" has little meaning. All of the modern races of mankind are of mixed origin; the differences between them in this respect are merely in the extent of mixing that has occurred in recent times. The genus *Homo* is exceptional among all higher organisms in that it has undergone a phenomenally rapid evolution in terms of progress toward a new adaptive complex, and toward a tremendously increased dominance over its environment, without the accompaniment of any permanent adaptive radiation.

This consideration of the biological evolution of mankind can be concluded by some speculations about the kinds of selective forces which guided it. The most significant changes which these changes brought about were the following:

1. The change from the four footed gait of terrestrial apes and monkeys to the bipedal gait of man. This required a considerable change in the structure of man's skeleton, which can be traced whenever fossils are found which include the right parts.
2. The perfection of the hand for tool making. The generalized apes from which the line leading toward man diverged had already evolved hands with opposable thumbs, probably in connection with grasping branches of trees.
3. Increase in brain size and intelligence. This involved not only mere increase in size of the brain, but also particular development of those centers in which intelligent responses are localized. When whole skulls are available, these changes can be followed to some degree.
4. Change in the diet from fruits, hard nuts, and tough roots to softer food, including an increasing dependence on meat. This involved decrease in size of the tearing canines, the development of more regular surfaces on the grinding molars, and later the reduction of the size of the molars themselves. These changes are easier to trace through the fossil record than are any others, since teeth are preserved as fossils more often than are any other parts of the skeleton.
5. Increase in the ability to communicate with others, and to develop organized community behavior. Direct evidence for this type of change cannot be obtained from the fossil record, but it can be approached indirectly in various ways.

The greatest value to man of walking on his hind feet is the freedom which this posture offers his hands for holding tools or other objects, throwing them, or catching them. Since the first known ancestors of man who walked erect, the australopithecines, were tool users and probably obtained much of their food with the aid of them, the gradual change to the erect habit probably accompanied the increasing use of tools. Both of these changes were probably promoted by the change from life in the trees to existence on the ground, which began during the radiation of the dryopithecine stock in the Miocene epoch. Chimpanzees, man's nearest living relatives, often use tools such as sticks and stones. In their native habitat, they have recently been seen to break sticks or pieces of vine of the correct length for getting termites out of their nests and to remove the branches from these sticks in preparation for use. Observers of groups of chimpanzees saw young animals learning this simple art from their elders. When molested by baboons, chimpanzees often pick up stones to throw at them.

Since chimpanzees live most of their lives in trees and subsist on fruit, tools are of little use to them for their principal tasks of food getting. But for a ground feeding ape with similar intelligence the situation would be different. It might learn to use sticks for digging up nutritious roots or extracting rodents and ground-inhabiting insects from their burrows. Its need for throwing stones to frighten off enemies would be much greater than in the case of an arboreal ape. Once throwing stones for defense had become a regular habit, the ape might achieve enough accuracy to kill small game at a distance. Naturally, the more quickly the ape could stand erect to achieve this purpose, the more accuracy he would gain. There is good reason, therefore, to suppose that the use of primitive tools and the erect posture were both acquired gradually and in relation to each other. They probably began with the first tendency to spend long periods of time out of the trees. Based upon the principle of double function during a period of transfer, as discussed in the last chapter, we would expect that in the earliest stages of this transition the apes lived partly on fruit and retired to the trees for protection, making increasingly greater excursions into open territory in order to increase their food supply. A later stage in the transition could have been one in which the daytime hours were spent on the ground, but the apes retired at night to the trees for protection. This way of life would have been compatible with existence in the open savannas, where some of the African dryopithecines lived.

Since the brain of the tool-using australopithecines is not much larger than that of anthropoids, we might logically conclude that most of man's intelligence was acquired after he had become a regular tool user and walked erect. Intelligence, however, depends not only on the size of the brain in relation to that of the body, but also on the development of specific areas in the brain. Perhaps the australopithecines were considerably more intelligent than anthropoids because of the development of these centers. Even if the australopithecines were little more intelligent than anthropoids in their overall reactions to stimuli, the ability to imitate and learn must have much more highly developed in australopithecines than in typical anthropoids.

Nevertheless, the most rapid increase in brain size and presumably in intelligence took place during the evolution of *Homo erectus* from the australopithecines. This increase accompanied the development of well fashioned tools of a relatively constant design. At about the same time, men learned how to tame and use fire. With better tools, these primitive ape-men learned to hunt large game, presumably in cooperating bands. The selective advantage of more efficient communication under these conditions is obvious. Consequently, we may reasonably suppose that the period when man's brain was increasing most rapidly coincided with the evolution of his ability to invent and use language. We cannot overestimate the importance of speech to man's way of life. Not only is it essential for carrying out complex hunting maneuvers and teaching the art of tool making, but it is also basic to the development of ideas and plans for the future. If we wish to single out any one period in human evolution when our evolutionary line acquired the human state, we must point to the transition from australopithecines to *Homo erectus,* when the first primitive forms of speech probably evolved.

This probable increase in brain size from australopithecines to *Homo erectus* could be regarded as a typical example of directive selection, as this process was characterized in Chapter 4. As such, it deserves special attention in order to emphasize the point that this kind of selection can take place at rates which paleontologists regard as very rapid by relying on the variability present in any normal gene pool, and without the need for invoking an increase in mutation rate or any other stimulating factor except an increased intensity of selection. The mean capacity of the brain in australopithecines was about 500 cc, of *Homo habilis* 680 cc, and that of *Homo erectus* 1000 cc. The dating of available fossils indicates that *Homo habilis* possessed this mean brain size of 680 cc about 900,000 years ago, and the mean given for *Homo erectus* was reached about 600,000 years ago. This would indicate that the size of the brain increased about 47 per cent in 300,000 years. During the last 600,000 years, the brain capacity of man has increased about 45 per cent, suggesting that the rate of increase has slackened during this later period of human evolution.

From the point of view of the fossil record as a whole, this rate of advance is spectacularly rapid. When all but the most recent geological epochs are considered, a change which takes place over a million years is so rapid that it is regarded as being almost instantaneous (Figure 8-4). It is well within the margin of error inherent in dating any fossil having an age of 50 million years or more.

Looking in the other direction, however, this order of change is remarkably slow from the genetic point of view. The geneticist thinks in terms of the amount of variation existing in a population in each generation, and the change in the mean value for a character which is produced by selection from one generation to another. When we express in these terms the evolutionary changes in size of the human brain, we obtain the following results. In modern man, the differences in size between the largest and the smallest brains found in any one population is about 16 per cent of the mean size. Assuming that the range of variation in a population was about the same in earlier species, this would lead us to conclude that the difference between the largest and smallest brains in a population of *Homo erectus* was about 160 cc, and in *H. habilis* about 109 cc. Now the increase in mean size from that of *H. habilis* to the later *H. erectus* was 320 cc in 300,000 years. Even allowing for errors and for changes in the rate of evolution, an increase of 200 cc during a period of 100,000 years is the most rapid we could ever expect to find. What does this mean in terms of the amount of increase in the mean value per generation, compared to the variability existing in one generation?

Assuming that the average human generation is 25 years, the period of 100,000 years represents 4000 generations. Consequently, an increase in brain size of 200 cc during this period means that the mean size has increased only

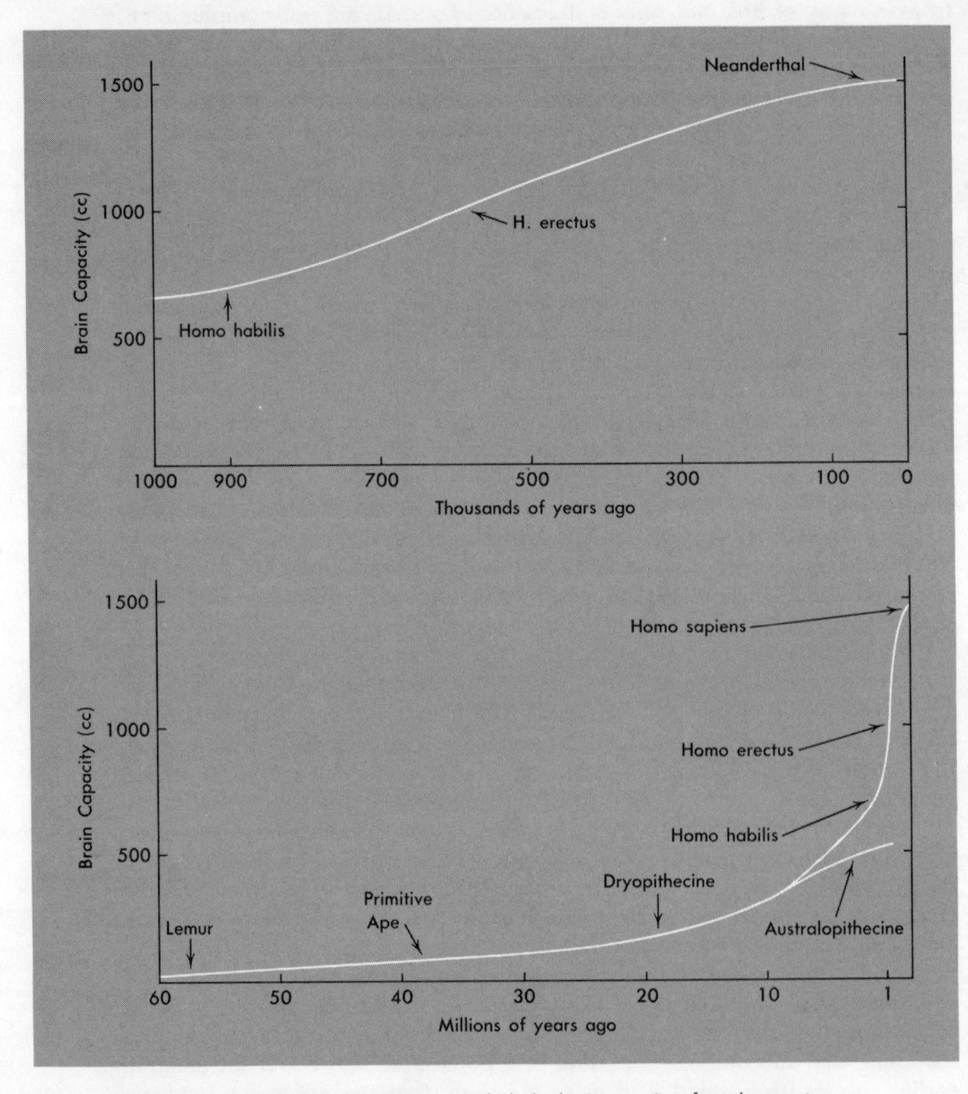

Fig. 8-4. Diagrams showing the rate at which the brain capacity of man's ancestors evolved toward increasing size. *Below,* the changes which took place during the entire course of primate evolution, from the beginning of the Tertiary Period to the present. The time span is about 60 million years. On this scale the increase in brain capacity appears very slow during almost all of this long period but very rapid during the last million years when the ape-like Australopithecines gave rise to modern man. *Above,* the curve of increase in brain capacity which appears when the last million years are stretched over a horizontal scale which is sixty times as great as that of the lower diagram. The "very rapid" increase in brain capacity now appears like an increase at a moderate rate. It probably was somewhat faster at about the time of the Java and Peking men (*Homo erectus*) than either before or since.

0.05 cc per generation! At this rate, the amount of increase per generation is only a tiny fraction, less than $\frac{1}{20}$ of one per cent, of the variation present in any one generation (Figure 8-5). It is many times smaller than the increase in mean value for quantitative characteristics which have been subjected to artificial selection by breeders working with crop plants and domestic animals. These calculations demonstrate dramatically what slow rates of response to natural selection are needed to bring about even the most rapid evolution which is recorded by sequences of fossils.

From the stage of *Homo erectus* to that of modern man, the greatest selective pressures were for developing further the centers of intelligence. During at least the past 40,000 years there has been little selective pressure for changed anatomical characteristics. This does not mean, however, that natural selection has ceased to operate. The drastic changes which have taken place in man's way of life during this period have subjected him to entirely new selective pressures. Crowding in cities increases dirt and disease, so that until very recently, a high premium was placed upon genes for disease resistance. The emotional stresses of life in large, highly organized

Fig. 8-5. Diagram showing how the most rapid phase of evolution of the hominid brain appears when represented in terms of human generations (assuming 25 years per generation). The mean increase in brain capacity per generation required to produce the maximum slope of the two previous curves is 0.057cc. This is insignificant compared to the amount of variation between individuals of a normal human population, which in modern man is about 16 per cent of the mean value. In the diagram, the sloping line expresses the amount of increase in mean capacity over 400 generations which would be required to produce the slope of the curve represented in figure 8-1, and the vertical bars, placed at intervals of 50 generations, represent the amount of variation per generation which would be expected, based upon that found in modern human populations.

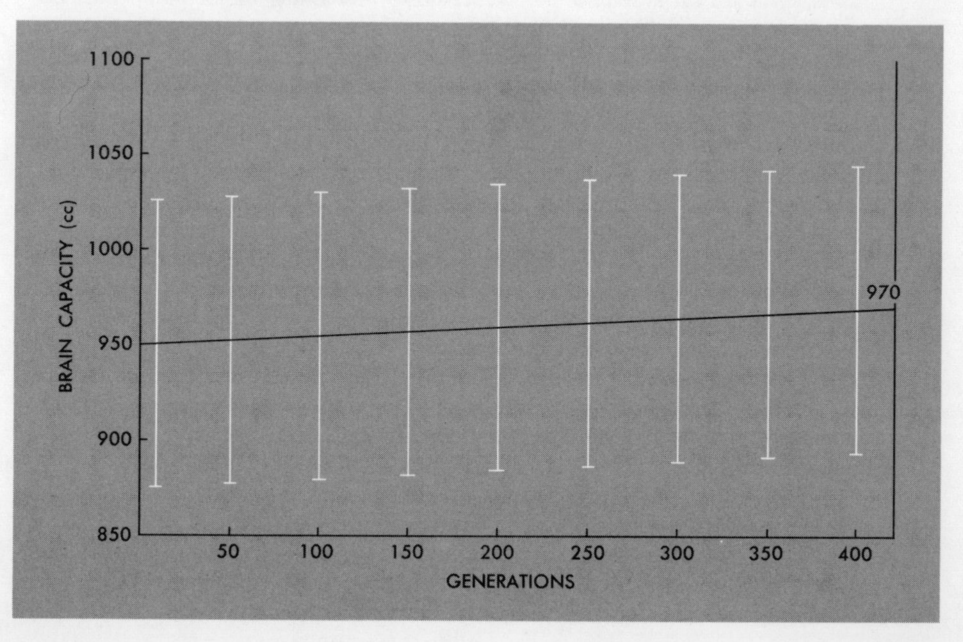

communities are very different from and much greater than those accompanying life in small hunting communities such as existed when man first appeared. In response to these stresses, selection must now be operating in much more subtle but equally important ways. Analyses of these kinds of stresses, however, belong to the tasks of the psychologist; a biologist is not qualified to discuss them.

ORGANIC EVOLUTION AND CULTURAL EVOLUTION

Even though the anatomical characteristics of man have changed little in 40,000 years, this does not mean that evolution has come to a halt or has materially slowed down. If we define evolution as an orderly sequence of changes in the relationships between the population and its environment, exactly the reverse has been the case. In these terms, the human species has been and still is evolving at an exceptionally rapid pace, and in an entirely new direction from that which any species population has previously taken. The earliest men which were anatomically like ourselves were living in caves, possessed only the crudest of clothing, ate only simple foods, and probably suffered often from a scarcity of even these. Their numbers were tiny compared to the enormous and ever increasing population of contemporary man. They had continually to defend themselves against animal predators, and were at the mercy of extreme changes in their physical, inanimate environment, such as floods, sudden storms, drought, and famine. Modern, civilized man has gained control over both his inanimate environment and the animals which formerly were his living predators. He has provided the means by which our species could live on this earth in peace, safety, and prosperity for an indefinite period. His only remaining major problem is how to adjust his own society so that his potential ways of achieving his goals can become a reality. The only major enemy of contemporary man is mankind himself. Such a situation has never before existed on this planet.

The way in which this dominance has been achieved is, however, so different from organic evolution as it has taken place in all populations of organisms except for modern man, that it must be given a different name and defined in a different way. We can call it CULTURAL EVOLUTION. Cultural evolution resembles organic evolution in that both processes result in progressive changes in the relationships between the evolving population and its environment. Both, moreover, depend upon the presence of a store of variability within the population, and upon processes which can direct this variability into particular channels. Cultural evolution, like organic evolution, can progress in many different directions, can reverse itself temporarily in some respects, can progress rapidly or slowly, and can remain constant at a stabilized equilibrium for long periods of time. Here, however, the resemblance stops.

The stored variability which makes cultural evolution possible is not genetic in nature, but consists of ideas, inventions, traditions, laws, customs, and all of the other learned responses by which society is regulated. Individual men acquire their share of it not through heredity but through conditioning, training, and learning. The store of variability is increased in size not by means of genetic change but by additions to our store of inventions, ideas, laws, and customs. Direction is given to cultural evolution not by selection which depends upon death or differential reproduction of individuals, but by imitation, conscious selection, foresight, and planning.

Another profound difference between organic and cultural evolution is the way in which changes in the relationship between population and environment take place. Alfred Russell Wallace, who conceived of the idea of natural selection almost at the same time as Darwin (see Chapter 1), first made this aphorism: Animals evolve by adjusting themselves to their environment; man by adjusting the environment to himself. From the biological standpoint, man is anatomically one of the most unspecialized of higher animals, and without culture we are poorly equipped to cope with our environment. We have no claws, our teeth and jaws are weak, and we cannot spring upon our prey. But we have learned to domesticate our food animals so that they are always at hand, to kill them easily and mercifully, and to prepare them according to a variety of highly developed arts and skills which have converted feeding from a routine necessity to one of the greatest and most refined pleasures of modern life. Man's ancestors obtained plant food by plucking, gathering, and digging. Modern man has, for this purpose, transformed most of the habitable surface of the earth from primeval forest and prairie into a vast, neat parkland especially designed for his own use. By consciously guiding their evolution, he has created a number of new varieties and species of food plants particularly adapted to his needs.

Men cannot run very rapidly compared to other animals, their ability to swim is incomparably poorer than that of any aquatic animal, and they cannot fly at all. Yet we have invented machines by which we can propel ourselves over the earth's surface, over and through the water, through the air, and into outer space at speeds which were inconceivable until modern times. Our greatest problems occur, as every commuter and Sunday tripper knows, when we bring too many machines into one place at the same time.

A study of cultural evolution is obviously outside of the field of biology. This discussion must end, therefore, now that the writer has pointed out the principal differences between the two kinds of evolution. Their intensive study in mankind will require cooperation between biologists, anthropologists, psychologists, and sociologists. Such studies should be one of the most important disciplines of higher learning in the future.

EVOLUTION IN THE FUTURE

Although gazing into the crystal ball is not usually regarded as a legitimate occupation for a scientist, perhaps the reader will permit me at this point to step out of my scientific role and indulge in a bit of it. As I see it, evolution in the future is destined to be dominated by the cultural evolution of mankind. To an increasing degree, other animals and plants will spread and evolve, become extinct, or remain stagnant, either according to the will of mankind, or because they can take advantage of the environments which man has created without being checked by him. The evolution of man himself will continue to be dominated by cultural evolution, with organic evolution assuming an increasingly subordinate role. I agree with the late Père Teilhard de Chardin, an eminent paleontologist, philosopher, and poetic prophet of human destiny. He states in his book, *The Phenomenon of Man,* that at this very instant of evolutionary time we stand at the boundary line between the biosphere, the realm of life and organic evolution; and the noosphere, the realm of the human mind and the evolution which it will generate in the future.

This does not mean that in the future the human species will regress or atrophy in respect to its bodily features. On the other hand, we have a number of built-in selective forces which are acting to maintain the human body within the normal range of anatomical characteristics existing at present. One of these is the series of complex objects and machines which we have built, from beds to houses, motor cars, the interiors of airplanes, and factory benches, all of which are built to accommodate men and women having a definite range of size and bodily structure. Another is the increasing interest of men in sports and outdoor living as a means of recreation, for which a well and normally constructed body is essential. Last and by no means least is conscious selection of mates. Handsome men and beautiful women are those which have normal features and a standardized, relatively narrow range of bodily measurements. It is true that beautiful women do not, on the average, have the most children, largely because of competing interests on stage, screen, and promiscuous manhunting.. This, however, is fortunate. If there were a high correlation between standardized beauty and fecundity, our gene pool would be harmfully reduced in size, and many of the most precious gene combinations, found in some "far out" people, might be lost. On the other hand, homely men and women, who in bodily features deviate a long way from the norm, have a lower chance of securing mates and producing children than do more normal people. Under most conditions of civilization, our voluntary method of choosing mates is producing a healthy balance between, on the one hand, too great standardization and reduction of the gene pool and, on the other hand, evolution in directions which might lead the human population toward an even greater disharmony between its biological characteristics and the demands of cultural evolution than now exists.

This balance is obviously a form of stabilizing selection. I expect, therefore, that interactions between human populations and their future environments will continue to be of a stabilizing nature with respect to their bodily structure. If people are still on the earth a million, ten million, or even a hundred million years from now, they may continue to look very much like ourselves, just as modern opossums, lungfishes, and king crabs are hardly distinguishable from their ancestors which flourished long ago.

A final question which we might ask is: Can man direct his future evolution? The discussion presented in the last section should have made obvious to the reader the fact that, since much of the evolution in the future will be cultural rather than organic, a biologist is not equipped by his knowledge and training to attempt an answer to this question, and should not try to do so. I should like to conclude this book, however, by pointing out some biological aspects of the problem.

In the first place, no amount of guidance which we might attempt to give to the evolutionary process will be of any use to mankind until he has learned how to control the explosive increase of his population. As many authors have pointed out, our present rate of increase, if it continues unchecked for a few more generations, will produce a condition of "standing room only." Not only will there be a shortage of food even if the most optimistic hopes of plant breeders and agriculturalists for increasing our food supply are realized; there will not be enough space to house people in those parts of the earth where they can expect to live in reasonable comfort. Biological inventions for artificially controlling the birth rate are rapidly being improved. If we should devote to this problem even half of the ingenuity, manpower, and money that we are devoting to solving such problems as heart disease and cancer, its biological aspects could probably be solved relatively quickly. Again, the block to progress is sociological; the greatest enemy of man is mankind himself.

An evolutionary biologist must point out the fact that artificial control of the birth rate is merely a continuation of one of the dominant trends of evolution which can be seen throughout the animal kingdom. It is the natural law of organic evolution converted to the needs of man's cultural evolution. In all species of nesting birds, the number of eggs which a female lays and the number of young which she hatches are carefully regulated by natural selection to be in harmony with the amount of food which the parents can bring to the nest. In nearly all evolutionary lines of mammals, including that which led to *Homo sapiens,* one of the major trends of evolutionary advancement has been reduction in the birth rate combined with greater care of the young. It is imperative that we take all possible steps to continue this trend.

Secondly, this evolutionist believes that little will be gained by trying to breed a race of intellectual supermen. For perhaps five hundred thousand years, natural selection in man's evolutionary lineage has operated to reduce the rate at which the size of his brain has increased. Although brain size and intelligence are only roughly correlated with each other, we have good reasons for believing that the rate of increase in intelligence has also diminished since the evolution of *Homo sapiens.* Certainly, the ancient Egyptians who lived 5000 years ago and built the pyramids must have been every bit as intelligent as ourselves. If, therefore, natural selection has been operating to keep man's intelligence in a harmonious balance with his other characteristics, we must think twice before tampering with this trend. Specifically, society has a great need for individual men of genius and for great leaders, but such people do not always get along with each other easily. Every opera director knows that one cannot make a chorus out of prima donnas and star tenors.

One might argue that our efforts should be directed toward breeding people who are more able to cope with society and to adjust themselves to each other. At present, however, this is an impossible task. Neither biologists nor psychologists have the slightest idea of what genetic tendencies would promote these qualities the most. Even if they did, selection for these qualities could not be carried out under any of the conditions which now prevail in society, and the amount of regimentation which would be needed to make this possible would be highly repugnant to most people. Perhaps when we have come somewhat nearer to making the most of our opportunities for improving mankind through conditioning, training, and teaching the younger generation under optimal conditions for their social development, we can then consider whether anything can be done about the genetic aspects of social improvement.

On a more modest scale, human geneticists and evolutionists have been wrestling with the problem of how to reverse the frequently observed tendency of the poorer classes, who appear to be lower in their intellectual endowment, to produce more offspring than do those in higher stations of life, and presumably with greater intellectual capacities. In my opinion, this problem has often been exaggerated. We do not know how great is the genetic component of the difference in intellect between a ditch digger and a surgeon, a great musician, or a leading scientist. Since, however, the genetic component of intelligence consists of very many separate genic elements, and since the human gene pool is very generalized and heterogeneous, we can be reasonably sure that, just as every private in Napoleon's army carried a marshal's baton in his knapsack, every ditchdigger in the United States carries in his germ cells some genes which could contribute toward making a Kennedy, a Shakespeare, a Beethoven, or an Einstein. If this is so, then our problem again becomes sociological. We must create in our society a climate which will foster voluntary adjustments of the birth rate in all classes to meet the needs of society.

The multitudinous problems of both organic and cultural evolution are by no means solved. For further progress toward their solution, cooperation will be needed between scientists of many disciplines, between natural scientists and social scientists, as well as between those who approach humanity from a more mundane point of view and those who, as leaders of the world's religions, are seeking to improve the spiritual qualities of mankind. This author hopes that his contribution has, in at least a small way, helped to further this end.

Chapter Summary

Recent studies of both morphology and comparative chemistry of proteins have shown that *Homo sapiens,* chimpanzee, and gorilla are more closely related to each other than any of these species is to the other anthropoid apes, orangutan and gibbon. All three of the first mentioned species are probably descended from a group of apes which was common in Eurasia and Africa during the Miocene epoch. The immediate ancestors of *Homo* were the australopithecines, which lived in North Africa and Eurasia at the end of the Pliocene and the beginning of the Pleistocene epoch. The earliest species of *Homo, H. erectus,* was widespread to Eurasia during the middle of the Pleistocene, and probably evolved into modern man by a series of stages, without splitting into separate species. The different characteristics which distinguish man from the apes probably evolved at somewhat different rates. One of the earliest human characteristics to appear was the use of tools, which preceded the increase in brain size and accompanied the change from four footed gait to erect posture. The increase in brain size during the early part of the Pleistocene epoch, though very rapid from the paleontologists' point of view, was probably very slow in terms of alteration in mean brain size per generation.

The most distinctive feature of modern man has been cultural evolution, which has taken place with very little accompanying change in visible characteristics of man's body. It has involved primarily the modification of his environment to suit his needs, rather than hereditary modifications of the body to suit his environment, as was characteristic of evolution in all other species of animals. The processes of cultural evolution are not controlled by the biological processes of mutation, gene recombination, and natural selection; but by invention, development of ideas, learning, foresight, cultural diffusion, and conscious selection. These facts must be regarded as basic to any prophecies about the future of evolution, or any attempts to control it.

Questions for Thought and Discussion

1. What reasons can you suggest for the apparent absence of speciation in the recent evolutionary history of mankind? Do you think that evolution in the future will be promoted if attempts are made to divide man artificially into reproductively isolated populations, which might evolve into separate species?

2. Discuss possible relationships between tool using, the erect posture, and the increase of human intelligence. How may we obtain better information to help us understand the relationships between these trends?

3. Give arguments for and against the use of the words *cultural evolution* to characterize the kinds of changes which man has undergone during the past 40,000 years.

4. Make your own prognostications regarding the future of mankind, and discuss them with your fellow students and friends.

SUGGESTED READINGS

A very large number of books have been written about evolution, from every possible point of view. Since the author assumes that the reader of this volume will be most interested in learning more about processes of evolution than about the evolutionary history of the various forms of life, the suggestions made below have this point in mind.

General references

Darwin, C. *The Origin of Species.* 1st ed., London, 1859. Facsimile with introduction and bibliography by Ernst Mayr, Harvard University Press, 1964. Sixth London Edition, 1872. This is the edition which has been reprinted in numerous forms and is most widely read.

Dobzhansky, T. *Genetics and the Origin of Species,* 3rd ed., New York: Columbia University Press. 1951.

Grant, V. *The Origin of Adaptations.* New York: Columbia University Press. 1963.

Mayr, E. *Animal Species and Evolution.* Cambridge, Mass.: Harvard University Press. 1963.

Simpson, G. G. *The Major Features of Evolution.* New York: Columbia University Press. 1953.

Tax, S., ed, *Evolution After Darwin: The University of Chicago Centennial.* Vol. 1, *The Evolution of Life.* Chicago, Ill.: Chicago University Press, 1960.

Chapter 1.

Carter, G. S. *A Hundred Years of Evolution.* New York: The Macmillan Company, 1957.

Chapter 2.

Bonner, D. M. and S. E. Mills, *Heredity,* 2nd ed., Englewood Cliffs, New Jersey: Prentice-Hall, Inc., 1964.

Chapter 3.

Darlington, C. D. *The Evolution of Genetic Systems,* 2nd ed. New York: Cambridge University Press. 1958.

Grant V. *The Architecture of the Germ Plasm.* New York: John Wiley & Sons, Inc., 1964.

Chapter 4.

Ford, E. B. *Ecological Genetics.* New York: John Wiley & Sons, Inc., 1964.

Lerner, I. M. *The Genetic Basis of Selection.* New York: John Wiley & Sons, Inc., 1958.

Wallace, B. and A. Srb, *Adaptations,* 2nd ed. Englewood Cliffs, New Jersey: Prentice-Hall, Inc., 1964.

Chapter 5.

Clausen, J. *Stages in the Evolution of Plant Species.* Ithaca, N.Y.: Cornell University Press. 1951.

Lack, D. *Darwin's Finches.* New York: Cambridge University Press. 1947.

Chapter 6.

Anderson, E. *Introgressive Hybridization.* New York: John Wiley & Sons, Inc., 1949.

Stebbins, G. L. *Variation and Evolution in Plants.* New York: Columbia University Press. 1950.

Stebbins, G. L. The Role of Hybridization in Evolution. Proc. American Philosophical Society 103:231–251.

Chapter 7.

Colbert, E. H. *The Evolution of the Vertebrates.* New York: John Wiley & Sons, Inc., 1955.

Romer, A. S. *The Vertebrate Story.* Fourth Edition. Chicago, Ill.: Chicago University Press. 1959.

Chapter 8.

Dobzhansky, T. *Mankind Evolving.* New Haven, Conn.: Yale University Press. 1962.

Tax, S., ed. *The Evolution of Life.* Vol. 2. *The Evolution of Man,* Chicago, Ill.: Chicago University Press. 1960.

index

a

b

c

d

g

h

m

n

o

p

q

r

s

t

u

v

w

x

z